THE PRINCIPLE OF SUSTAINABILITY

For Johannes, Christian and Olivia

The Principle of Sustainability
Transforming Law and Governance

KLAUS BOSSELMANN
University of Auckland, New Zealand

ASHGATE

Published by

Ashgate Publishing Limited
Gower House
Croft Road
Aldershot
Hampshire GU11 3HR
England

Ashgate Publishing Company
Suite 420
101 Cherry Street
Burlington, VT 05401-4405
USA

www.ashgate.com

British Library Cataloguing in Publication Data
Bosselmann, Klaus
   The principle of sustainability : transforming law and
   governance
   1. Environmental policy 2. Environmental ethics
   3. Environmental justice 4. Sustainable development
   I. Title
   363.7

Library of Congress Cataloging-in-Publication Data
Bosselmann, Klaus.
   The principle of sustainability : transforming law and governance / by Klaus
Bosselmann.
      p. cm.
   Includes bibliographical references and index.
   ISBN 978-0-7546-7355-2
   1. Sustainable development--Law and legislation. 2. Environmental law. 3.
Environmental ethics. I. Title.

   K3478.B67 2008
   344.04'6--dc22

                                                                    2008014372

ISBN 978-0-7546-7355-2

Mixed Sources
Product group from well-managed forests and other controlled sources
www.fsc.org Cert no. SA-COC-1565
© 1996 Forest Stewardship Council
FSC

Printed and bound in Great Britain by
MPG Books Ltd, Bodmin, Cornwall.

Contents

Foreword

In an age in which we are denuding the resources of the planet as never before and endangering the very future of humanity, sustainability is the key to human survival.

It is a concept which needs to be nourished from every discipline, every culture and every tradition. Unfortunately these have not been adequately used in strengthening and developing this concept. Professor Bosselmann's book provides a many-faceted approach to the concept which will give some idea of its conceptual richness and practical potential.

The concept is traced to its historical and philosophical foundations and is approached from a multicultural and interdisciplinary perspective. It is examined from the standpoint of international law, of which it is now an integral part and, at the same time, an attempt is made to stir an interest in the minds of average citizens on whom, in the last resort, its effectiveness depends. Indeed the book has a richness of historical, jurisprudential, anthropological, sociological, political, scientific and philosophical perspectives which make it rewarding reading.

The entire concept is permeated by the notion of ecological justice and this requires a balancing of the interests involved. Development is important, but not at the expense of sustainability. We owe an obligation to future generations who, under current legal systems, are unable to assert their rights and consequently remain voiceless when their rightful inheritance is being plundered by the present generation. All who permit this to happen must share in the blame. This inertness of the general public, while their children and their children's children are so inequitably treated, is due to lack of communication of the perspectives so eminently dealt with in this book. It deserves to reach as wide an audience as possible.

The important area of the linkage between human rights and ecology is exhaustively dealt with. Human rights doctrine is intensely relevant to environmental protection though not often perceived and this volume does much to bring home its practical importance in this field.

The concept of trusteeship, as opposed to the traditional belief of dominance over our environment, needs also to be widely publicized and this book is an important vehicle for doing so.

The duties of states to promote the concept of sustainability are another greatly neglected area, for decision-makers pursuing short-term objectives tend to neglect the long-term perspectives. This is a principle that needs to be firmly embedded into the structure of contemporary international law, and this volume is a valuable contribution to this end.

All who read it will be the wiser for the experience. It has a message for everyone in terms within the reach of everyone. It is an important contribution towards achieving the more caring world we all desire to see established.

Judge C.G. Weeramantry
Former Vice President of the International Court of Justice
Emeritus Professor of Law, Monash University

Acknowledgements

When the forms of an old culture are dying, the new culture is created by a few people who are not afraid to be insecure.

Rudolf Bahro

It is appropriate to first thank Rudolf Bahro, the Green philosopher and pioneer of a new culture. During his years at the Humboldt University Berlin (until his death in 1997) I had the privilege of working with him, in various ways, in the field of social ecology. I learned how ecology must inform social sciences and jurisprudence and, even more so, how flawed the divide between sciences and humanities really is. Humanity will not progress in its – increasingly desperate – search for sustainability, if we continue to separate scientific 'facts' from ethical 'values'. Finding facts is value-based and defining values is fact-based, at least, when it comes to sustainability. The reality of one planetary ecosystem and how to live within its boundaries can only be grasped if we learn to think globally, holistically, and responsibly. Like the deep ecology philosopher Arne Naess, Rudolf Bahro had an eco-centric, not human-centred worldview. Unlike Naess, Bahro was steeped in political debate. He insisted on addressing the sustainability issue in a socio-economic, political and legal context. Essentially, this issue is one of justice, seeking a 'fair share' between rich and poor, presence and future, humans and nature. Bahro encouraged me to say things that needed to be said, but that I knew would not be well received, especially among politicians and law-makers.

Like most of us, I work in networks of deeply valued people. Often one's own ideas cannot be separated from the ideas of others, nor should they perhaps. I am blessed with an amazingly loving family and the fact that my wife Prue Taylor is also my closest friend and most influential colleague. Her courage to act ethically and speak the truth regardless of expediency is an ongoing inspiration. My colleagues at the University of Auckland Faculty of Law have been generous with their time and manifold support. Particular thanks are due to Paul Rishworth, Bruce Harris, David Williams, Jane Kelsey, Ken Palmer, David Grinlinton, Tim McBride and Nin Tomas. Much of what I have learned about sustainability stems from endless conversations with ethically trained scholars – whether environmental philosophers such as Laura Westra, Ron Engel, Nigel Dower, Franz-Theo Gottwald and Michael Hauskeller, environmental scientists such as Bill Rees, Colin Soskolne, Brendan Mackey and Jack Manno, economists such as Wayne Cartwright, Richard Westra and Peter Ulrich or environmental lawyers such as Ben Boer, Nick Robinson, Alexander Kiss, Dinah Shelton, Parvez Hassan, Sheila Abed, Antonio Benjamin, Ingo Sarlet, Ricardo Libel Waldman, Dire Tlady, Don Brown, Ben Richardson, Christina Voigt, Ulrich Beyerlin, Michael Schroeter, Hanfried Blume and, of course, Prue Taylor. All these and many other fine scholars and activists are members of global networks that helped pave the way towards the principle of sustainability: IUCN Commission on Environmental Law, its Ethics Specialist Group, Earth Charter International, IUCN Academy of Environmental Law, Global Ecological Integrity Group and the International Law Association with its Working Group on Sustainable Development.

For editorial assistance I am very grateful to Briar Charmley and Mia Koning. Finally, my sincere thanks to Alison Kirk from Ashgate for her support and guidance and Jenny Macgregor for her meticulous proofreading.

K.B.

Introduction

This book is about a most popular idea. It is so popular today that all sectors of society have embraced it. Even business and politics promise sustainability at every possible occasion. Living sustainably is an idea whose time has come. Or has it?

All good ideas take time and often they are misunderstood. In the case of sustainability, there has been a period of incubation followed by expert debate and eventual popularization through international politics. This has done more bad than good. The concept of 'sustainable development' lost its core meaning somewhere between the 1980s and today.

The idea and terminology of 'sustainable development' goes back to Robert Prescott-Allen, IUCN staff writer and principal author of the 1980 World Conservation Strategy. Section 13.1 reads: 'Ultimately the behavior of entire societies towards the biosphere must be transformed if the achievement of conservation objectives is to be assured. A new ethic, embracing plants and animals as well as people, which will enable human societies to live in harmony with the natural world on which they depend for survival and well-being'. The key words are 'ultimately ... transformed' and 'a new ethic'. A few years later, the World Commission on Environment and Development ('Brundtland Commission') also called for transformation and a new ethic. However, the popular notion became: 'Sustainable development is development that meets the needs of the present without compromising the ability of future generations to meet their own needs'. This vagueness opened up the possibility of downplaying sustainability. Hence, governments spread the message that we can have it all at the same time, i.e. economic growth, prospering societies and a healthy environment. No new ethic required.

This so-called weak version of sustainability is popular among governments and business, but profoundly wrong and not even weak as there is no alternative to preserving the Earth's ecological integrity. We can accept or dispute this insight, but should not confuse mere integration of economic, social and environmental policies with sustainable development. Integrating policies is an important step towards it, but only the first step.

The search for sustainable development is foremost an investigation into the truth of sustainability and I mean 'truth'. As a university professor it is my job to seek truth while allowing others to test how truthful my ideas may or may not be. I am convinced that there is truth behind the idea of sustainability that can be discovered in the same way as we have discovered the truth of evolution. It may well be that, in the future, we will have different explanations for living processes; I suggest however that Charles Darwin's theory of evolution has been one of humanity's greatest discoveries. I also believe that we can build sound ethics, law and governance around this discovery.

One of Darwin's most fervent supporters was Thomas H. Huxley, who famously said: 'Irrationally held truths may be more harmful than reasoned errors' ('Science and Culture', 1881). That human population and economy can continue to grow without threatening the sustainability of human life is an irrationally held truth. And while it may turn out be an error, it stands to reason that our survival depends on the ability to respect and maintain the Earth's ecological integrity. This is the core idea of the principle of sustainability.

On the day when the world's oldest and largest environmental organization, the World Conservation Union (IUCN) was founded in 1948, T.H. Huxley's grandson, the famous writer Aldous Huxley, wrote to his brother Julian Huxley, then Director General of UNESCO and one of the founders of IUCN:

> I come to feel more and more that no system of morals is adequate that does not include within the sphere of moral relationships, not only other human beings, but animals, plants and even things … If we don't do something about it pretty soon we shall find that, even if we escape atomic warfare, we shall destroy our civilization by destroying the cosmic capital on which we live. Our relationship with earth is not that of mutual beneficial symbiosis; we have become the kind of parasite that kills its host, even at risk of killing itself.

This letter is quoted at the beginning of the new IUCN Programme 2009–2012 'Shaping a Sustainable Future' which was adopted at the World Conservation Congress in Barcelona in October 2008. The World Conservation Union unites more than 1000 states, governmental agencies and international and national environmental organizations working together towards sustainability. The Programme advocates 'a new era of sustainable development thinking' that 'recognizes the intrinsic value of nature' and 'also recognizes that ecosystem health underpins human well-being'. The description of ecological sustainability is reflective of the Earth Charter that the OUCN had adopted in 2004 to guide its programme and policy development.

The Earth Charter is the 'universal declaration' and 'new charter' that the Brundtland Report over 20 years ago had called for. During the preparations of the 1992 Earth Summit in Rio, states had contemplated an Earth Charter, but, as mentioned, stayed away from the new ethic. It was, therefore, left to global civil society to develop an Earth Charter as an ethical framework for a just, sustainable and peaceful future. By defining respect and care for the community of life and ecological integrity as its two main principles, the Charter gives shape and meaning to the principle of sustainability. Adopted in 2000, i.e. 20 years after the World Conservation Strategy, the Earth Charter has brought the concept of sustainable development back to its original meaning.

Has the international community finally discovered the truth of sustainability that had been lost in the sustainable development debate? The fact that the Earth Charter has been endorsed by UNESCO, IUCN and a number of states and is recognized as the guiding document for the current United Nations Decade of Education for Sustainable Development (2005–2014) is significant. Quite obviously, the world is in search for sustainability and perhaps better informed than ever before. After all, we experience the impacts of unsustainable development on a daily and ever more dramatic basis.

If every good idea needs its time and opportunity to mature, now is the time. Sustainability can be defined as a fundamental principle of law and governance. It has reached a degree of maturity that allows for an examination of its meaning and legal status. This can be done in a similar way as other fundamental principles such as justice or freedom have been examined and promoted.

My own work in this area was greatly encouraged by efforts of the New Zealand government to shape legislation around the concept of sustainability. When I joined the Law Faculty of the University of Auckland in 1989, I found this intriguing statement of the Ministry for the Environment: 'Sustainability is a general concept and should be applied in law in much the same way as other general concepts such as liberty, equality and justice'. Is there a greater challenge for a legal scholar than exploring how it could be applied? If sustainability is likened to the foundational pillars of Western civilization what does that say about today's civilization? How civilized can a society be if it is not even capable of sustaining itself?

The former Vice President of the International Court of Justice, Christopher Weeramantry, is widely recognized as the leader of this important debate. In his Separate Opinion to the ICJ's 1997 Case Concerning the Gabçikovo-Nagymaros Project (*Hungary v. Slovakia*), for example, he described ecologically sustainable development as 'not merely a principle of modern international law. It is also one of the most ancient ideas in the human heritage'. In his judgment, Weeramantry traversed the ancient wisdom of irrigation and attendant legal systems of Sri Lanka, Africa, Iran, China, and of the Inca civilization. Reference was also made to the value systems of various cultures which reveal a universal love of nature, a desire for its preservation, and the 'need for human activity to respect the requisites for its maintenance and continuance'. One gets the distinct impression that modern legal systems are, in some important respect, more primitive than ancient civilizations. Perhaps there is wisdom in the fact that we acknowledge 'general principles of law recognized by civilized nations' (Statute of the International Court of Justice) as a source of international law, but no nation dared to recognize sustainability as a general principle. Civility has to be aspired to, earned and maintained. Anthropologists such as Oswald Spengler, Arnold Toynbee or Jared Diamond remind us how fragile civilizations have been.

If sustainability is foundational for the project of civilization, it deserves the full attention of everyone interested in continuing this project. Disturbingly, governments have not been forthcoming to take leadership or encourage research into much needed new strategies for policy development, law and governance. And as sustainability lies squarely across individual disciplines it easily slips through the gaps. There is neither a discipline focusing on sustainability issues nor an ethos of interdisciplinary research surrounding them. Obsessed with economy-related outcomes, most research institutions, including universities, are blind-eyed in this regard.

It largely falls upon civil society to develop the sustainability agenda, but that is perhaps the appropriate level. Civility – the central virtue of civilization – requires courage and constructive confrontation. And citizenship is the vehicle for political change. It is probably true that we are now experiencing what some have called a 'new green movement', an outburst of creativity in society, business and technology towards a low-carbon economy. There is a certain optimism of endlessly creative solutions and a quality of life that is both comfortable and energy saving. However, it does take a strong sense of citizenship to turn this new optimism into a catalyst for a change of institutions and governance.

The required changes involve all levels of society, but foremost a change of how the environmental problems are being dealt with politically. Environmental governance is still the poor cousin of economic governance, rendering the concept of sustainable development an unfulfilled promise. For sustainable development to become the overarching paradigm in law and governance, its conceptual core, i.e. the principle of sustainability, needs to be (re-)discovered, explained, defined and applied.

The thesis of this book is that sustainability has the historical, conceptual and ethical quality typical for a fundamental principle of law. Like the ideals of justice and human rights, sustainability can be seen as an ideal for civilization both at national and international level. When accepted as a recognized legal principle, sustainability informs the entire legal system, not just environmental laws or not just at the domestic level. While certain legal developments in individual countries such as New Zealand, the United States and Germany, and in the European Union as a whole, are of particular interest and will be discussed, the book does not take a comparative approach. This is not a study of comparative environmental law. Nor is international environmental law as such the subject of investigation. Rather the book investigates how sustainability informs universal principles used in domestic law as well as international law.

Taking a global perspective, the book aims to overcome the dichotomy between international law and municipal law. In the age of globalization, the separation between both spheres is withering somewhat, but the great divide between them, the concept of state sovereignty, remains the cornerstone of the world's legal heritage. Overcoming the dichotomy has certainly been the goal of international human rights law and, perhaps even more so, international environmental law. Human rights are universal in character and ought to be followed regardless of the world's cultural diversity. However, this is in dispute. Given the diversity of cultural traditions and sometimes rivalries between them, the principle of universally accepted human rights is, at best, only emerging. The environment, on the other hand, is global by nature and the functions of the Earth's ecological systems are felt everywhere beyond any cultural identity. The environment is the great unifier of humanity, at least in the sense of a shared concern. Environmental protection and, in fact, the principle of sustainability, are a truly global challenge. To this end, the book makes a contribution to a theory of global law.

One other way to describe the book's methodology is a study into the 'greening' of the fundamental principles of law and governance. If sustainability is a fundamental principle, as will be argued here, then it needs to inform other fundamental principles. The idea(s) of justice, for example, inform the concept of human rights. Likewise, human rights keep informing interpretations of distributional justice. The state is constrained by the guarantees of human rights and justice, but is also their guarantor. Mutual reinforcement of fundamental concepts has created the modern state as it functions today. It is reasonable, therefore, to ask how the challenge of protecting the global environment impacts on the modern state with its traditional functions and how the 'greening' of its governance, institutions and law can be articulated.

The first chapter traces the meaning of sustainability both as an idea and as a term. Going back into times preceding the modern debate on sustainable development we can discern certain characteristics not only in ancient cultures that Christopher Weeramantry referred to, but also in European history. The idea and even the term sustainability (German *Nachhaltigkeit*) have been understood and practised long before the modern debate of the 1980s. The fact that sustainability was a legal term with a defined content and used in legislation is important for the interpretation of the modern composite term 'sustainable development'. It would be wrong to assume that this construct only emerged following the Brundtland report and could only be interpreted accordingly. Moreover, sustainability has continued to evolve as a fundamental idea, not least through scientific and ethical reasoning. The Earth Charter, in particular, has been helpful in defining this idea with clear contours and meaning. Essentially, sustainability means maintenance of the integrity of the Earth's ecological systems.

The second chapter examines the legal status of sustainability. There is no commonly agreed typology of legal norms, but using the fairly established typology of international law we can describe sustainability as a legal principle. As such it creates meaning and legal status for the concept of sustainable development. The fact that there is a wide range of opinions on the legal quality of sustainable development is largely owed to its vagueness. The lack of definable content has long hampered, even hindered the political and legal acceptance of sustainable development. The principle of sustainability gives this concept more exact contours allowing its qualification as a principle of international law. However, the principle of sustainability reaches further. It is broad and fundamental like some other pillars of modern society, i.e. justice, equality and freedom. With its wider spatial and temporal dimension it impacts on the meaning of justice, equality and freedom.

The third chapter then explores the relationship between sustainability and justice. If humans are to be concerned with maintaining the integrity of the Earth's ecological systems they need to accept responsibility for non-human aspects of nature. This raises the issue of environmental ethics. There are various theories how environmental ethics could inform the idea of justice. Liberal concepts aim for reconciling care for non-humans and justice in a different manner than ecological concepts. Essentially, the contemporary debate surrounds John Rawls's theory of justice. Can his liberal theory be sustained by adding a 'green twist' (Wissenburg 1999) or do we need a genuine ecological theory of justice? The concept of ecological justice, it will be argued, is not only compatible with the principle of sustainability, its three elements (of intragenerational, intergenerational and interspecies justice) provide clear guidance for the interpretation of existing laws and the design of future laws.

Chapter 4 examines the relationship between sustainability and human rights. That human rights have an environmental dimension is well established in human rights theory and practice. Courts have for some time now acknowledged that environmental degradation threatens the enjoyment of rights to life and well-being, privacy, property and others. There is also wide recognition of new environmental rights, from procedural rights (to information, public participation, judicial review) to a distinct human right to a healthy environment. While these developments indicate a certain greening of human rights and better enforcement of environmental laws, they also tend to foster anthropocentric reductionism. From a sustainability perspective, human rights have to be complemented by responsibilities for the environment. This has been largely overlooked by the environmental rights discourse. Environmental responsibilities cannot be understood as mere moral duties or merely expressed in environmental laws. The fundamental character of human rights requires that they themselves need to incorporate such duties ('ecological limitations'). Only then will it be possible, for example, to abolish the right to pollute and exploit (expressed in property rights) and install a right for sustainable use.

The fifth chapter deals with the legal concept of the state. Given the pivotal role that the state has for protecting its citizens and natural environment, we need to ask whether it can, in fact, fulfill this role. To a degree, threats to citizens and their environment are global and not necessarily 'home-made'. While people, institutions and corporations are culprits, they are also victims in need of protection. The principle of sustainability, therefore, redefines the role and functions of the state by adding a fiduciary role. Internally, the constitution can provide for a state obligation to protect the environment (as a number of states have done). Externally, territorial sovereignty can be understood to include a stewardship or trusteeship function. As the state's territory is simultaneously national environment and part of the global environment, there is a need to complement the recognized permanent sovereignty over its own natural resources with a trusteeship role with respect to the global environment. The jurisprudential debate surrounding state sovereignty has made good progress in recent times with some commentators advocating the idea of the state as environmental trustee. Its practical consequences include an obligation to protect the global environment as a general principle of international law and the establishment of global institutions of trusteeship, but also vastly increased political and legal pressures on treaty negotiations, for example, with respect to climate change.

The final chapter is devoted to governance in an attempt to describe governance for sustainability. Here we first examine the current system of global environmental governance and efforts for its reform. The UN reform, for example, is a much publicized process with proposals for a revitalized Trusteeship Council, a global environmental agency, an international environmental court and other institutions to promote and review sustainable development. Outside the UN system, but often in cooperation with it, are international environmental organizations such as the IUCN representing the voice of civil society. The emergence of global civil society is among the most significant events in recent times. Hardly any activity or non-activity of states gets unnoticed by a critical international audience. Moreover, global networks of citizens and civil society groups reflect the advent of a new kind of citizenship. The concept of global citizenship, while purely ethical and not legal in nature, is of great political relevance. The traditional elements of national citizenship – identity and loyalty – are increasingly undermined (or enriched?) by global identity and loyalty. Constitutional law already recognizes multinational and supranational forms of citizenship, but what does it mean for governance when people define themselves as global and ecological citizens in addition to their national citizenship? State-centred governance models have yet to accommodate this new consciousness and confidence of citizens.

The principle of sustainability sets jurisprudence and law-making institutions on a new path. International law can no longer be perceived as a contractual arrangement purely between states and national law no longer as a purely domestic affair. Both levels penetrate each other. And as the global environment is perceived as our common home, legal instruments and principles will have to accommodate ecological citizenship. This 'new species' will insist on human rights and global justice, but also accept responsibility for its home called Earth.

Chapter 1

The Meaning of Sustainability

In this chapter I will argue that sustainability is a meaningful and powerful idea. The only reason why we may think otherwise would be that the term has been used in such a variety of meanings that it has become meaningless. Such criticism, I suggest, confuses the idea with the term. While the term may have been misused, the idea remains and continues to influence our thinking about the future.

What is Sustainability?

Sustainability is both simple and complex. Herein it is similar to the idea of justice. Most of us intuitively know when something is not 'just' or 'fair'. Similarly, most of us are fully aware of unsustainable things: waste, fossil fuels, polluting cars, unhealthy food, and so on. We can also assume that many people have a clear sense of justice and sustainability. For example, they feel that a just, sustainable world is desperately needed no matter how distant an ideal it may be.

In its most elementary form sustainability reflects pure necessity. The air that we breathe, the water that we drink, the soils that our food comes from are essential to our survival. The basic rule of human existence is to sustain the conditions life depends on. To this end, the idea of sustainability is simple.

But sustainability is also complex, again like justice. It is difficult to categorically say what justice is. There is no uniformly accepted definition. Justice cannot be defined without further reflection on its guiding criteria, values and principles. Such reflection is subjective by nature and open to debate. The same is true for sustainability. It cannot be defined without further reflection on values and principles. Thus, any discourse about sustainability is essentially an ethical discourse.

The term sustainability triggers a similar response to the term justice. Everybody agrees with it, but nobody seems to know much about it. We have only a vague idea what sustainability involves or how it could be achieved. We may be able to *imagine* a sustainable society, but probably not to how to get there. On the other hand, a 'just society' reflects an ideal which may never be fully achieved. Ideals such as justice, peace and sustainability are fundamental to any society. We cannot do without them.

Sustainability and justice evoke similar sentiments. In some ways, however, sustainability appears more distant than justice. There are several reasons for that. First, many of today's societies can be described as just, at least in a sense of providing the means for peaceful conflict resolution. By contrast, none of today's societies is sustainable. They are too deeply engrained in wasteful production and consumption to realize their unsustainable character. Second, the absence of justice is harder to bear than the absence of sustainability. Persistent unjust treatment of people by political regimes, for example, will not be tolerated for long. Either internal or external forces will revolt against it. Unsustainable treatment of the environment, on the other hand, is more likely to be tolerated. The reason is that people are less immediately affected by its impacts. The distance in space (global environment) and time (future generations) prevent us from acting with urgency.

Yet, perceiving sustainability with a similar immediacy as we perceive justice is entirely appropriate, precisely because the distances are vanishing. The world has become a small place and the future is already here. Climate change is an example in point. For a long time, the impacts of climate change appeared as distant possibilities. This is no longer the case. Now, climate change makes headlines on a daily basis. Since Al Gore's *Inconvenient Truth*, Nicholas Stern's report on the economic costs of global warming and George Bush's acceptance of climate change as a 'serious problem', the media have firmly embraced climate change as the most pressing issue of our time.

As we realize the impacts of climate change, we begin to feel its morality as possibly the biggest challenge. How can we justify the fact that our actions today will almost certainly threaten the planet's future? We are failing to meet the most basic obligation of each generation, i.e. to provide for the future of our children. This raises a moral question typical for sustainability *and* justice. How can we organize a fair distribution of goods and burdens throughout the generations?

It is hard to avoid the conclusion that sustainability fundamentally poses a challenge to the idea of justice. If a person lives at the expense of others, we consider this to be 'unfair'. If rich societies live at the expense of poor societies, we consider this also to be 'unfair'. Why then should it be acceptable to live at the expense of future generations and the natural environment? Whether or not sustainability requires, in fact, a rethinking of the idea of justice needs further consideration.[1] However, realizing the linkages between the two concepts also helps us to access the meaning of sustainability. It is an idea that refers to the continuity of human societies and nature.

Going back into history, we find that continuity of cultures and societies could only be ensured if ecological systems were sustained. Jared Diamond identified five factors contributing to the collapse of civilizations: climate change, hostile neighbours, trade partners, environmental problems and, finally, society's response to its environmental problems.[2] The first four may or may not prove crucial for the demise of society, Diamond claims, but the fifth always does. The salient point, of course, is that a society's response to environmental problems is completely within its control, which is not always true of the other factors. In other words, as his subtitle puts it, a society can '*choose* to fail'. The fact that choice is at the heart of continuity makes sustainability a matter of ethics. A society can choose to incorporate or to ignore the need to live within the boundaries of ecological sustainability.

---

1   See Chapter 3.

2   Diamond, J. (2005), *Collapse: How Societies Choose to Fail or Succeed* (New York, Viking Books).

It is at the level of basic values, therefore, where sustainability – like justice – needs to be conceived in the first place. For this reason, the vision of a 'just and sustainable society'[3] is not a distant dream, but conditional to any civilized society.

History gives us a clue why sustainability has always been a concern of society. The modern sustainability debate is by no means new, it only adopted the new focus on 'sustainable development'. Whether or not this focus has helped to understand the principle of sustainability or deviate from it is the big question.

The answer that will be offered in this chapter is that the concept of sustainable development is only meaningful if related to the core idea of ecological sustainability. We will see that sustainable development needs be understood as an application of the principle of sustainability, not the other way round. The vision of a 'sustainable society' is another, broader application of the same idea. Other applications can be seen in the terms 'sustainable growth', 'sustainable economy', 'sustainable production', 'sustainable trade' and so on. No matter how clear or confusing such terminological combinations are, they all employ a basic idea of sustainability.

With respect to 'sustainable development', the crucial question is how the concern for ecological sustainability is related to development, more precisely, the concern for prosperous development of people living today (intragenerational equity) and in the future (intergenerational justice). As will be shown, the sustainability debate since the Brundtland Report of 1987[4] has, to a large extent, overlooked the importance of defining these relationships. Sustainable development does not call for a balancing act between the needs of people living today and the needs of people living in the future, nor for a balancing act between economic, social and environmental needs. The notion of sustainable development, if words and their history have any meaning, is quite clear. It calls for development based on ecological sustainability in order to meet the needs of people living today and in the future. Understood in this way, the concept provides content and direction. It can be used in society and enforced through law. The legal quality of the concept of sustainable development firms up once its core idea is being realized.

A Short History of Sustainability

The meaning of sustainability can best be understood when we can ask whether there has ever been a sustainable society. If we interpret the Brundtland definition in a way that attributes equal importance to ecological, social and economic considerations, the benchmark for a sustainable society is extremely high. Was there ever equity between rich and poor, between sexes and ages, between countries and cultures and, at the same time, ecological sustainability and economic prosperity? Clearly, the answer is no. Pre-agricultural societies of hunters and gatherers have endured for

---

3   As, for example, expressed in the Earth Charter, available at <www.earthcharter.org>, accessed 1 December 2007.

4   World Commission on Environment and Development (1987), *Our Common Future*, 'Brundtland Report' (Oxford and New York, Oxford University Press), 65. Defines sustainable development as 'development that meets the needs of the present without compromising the ability of future generations to meet their own needs'.

a long time, the Australian Aboriginals, for example, for 60,000 years. Agricultural civilizations, like Ancient Egypt or the Indus valley, lasted for more than 5,000 years; however, from what we know, they were also shaped by inequity, oppression, violence and imbalances in all forms. If the characteristics of social and economic justice are part of the meaning of sustainability, then no society or civilization has ever been sustainable. Sustainability, in this sense, would remain a utopian idea, a distant goal that can never be achieved.

If, on the other hand, sustainability is brought back to its basics, the term becomes operable and meaningful. Before Brundtland, the term referred to a physical balance between human society and the natural environment. If the physical exchange processes between society and environment is upheld for a long period of time, a situation of sustainability can be observed. The question, whether societies have ever been sustainable, can be answered quite clearly and independently of whether they also have been 'just' or peaceful. So, what are the historical roots of sustainability and why should this matter to us today?

The Basic Idea

The idea of sustainability has its roots in the history of humankind. The Prince of Wales linked it even to the essence of humanity: 'deep within our human spirit there is an innate ability to live sustainably with nature'.[5] The 'innate ability' may refer to unfulfilled desires rather than actual abilities, however, the notion reminds us of our co-evolution with life as a whole. The desire for living in harmony with nature is undoubtedly part of our evolutionary heritage. But is the opposite not true as well? Considering the destructive forces of global corporatism and consumerism, we may think of harmonic relationships as a distant dream of the past. Yet, we can equally ask whether total global consumerism is any more 'innate' to human conditioning than wanting to live sustainably. Human existence has always been embedded in natural cycles and whether we realize it at the present time or not this will not change.

Christopher Weeramantry made some revealing observations in his separate opinion in the ICJ *Gabçikovo-Nagymaros* case 1997.[6] Referring to ancient agricultural and legal systems in the Americas, Africa, the Middle East and Asia, he recalls the long tradition of living in harmony with nature. Ancient civilizations were grounded in value systems that did not separate the human sphere from the natural sphere. It was inconceivable, for example, for North American Indians or Maya and Inca civilizations to seek economic prosperity at the cost of ecological sustainability. Exploitation and preservation were held together by the 'need for human activity to respect the requisites for its maintenance and continuance'.[7] Considering today's ecological crisis, Weeramantry suggests that the 'formalism' of modern legal

---

5   Foreword by HRH The Prince of Wales in Pye-Smith, C. and Feyerabend, G.B. (1994), *The Wealth of Communities: Stories of Success in Environment Management* (London, Earthscan), VII.

6   *Case Concerning the Gabçikovo-Nagmaros Project (Hungry/Slovakia)*, 1997 ICJ, 37 ILM 162 (1998), Separate Opinion of Vice-President Weeramantry.

7   Ibid. 18.

systems, deprived of such values and principles, must be overcome: 'The time has come when [modern legal systems] must once more be integrated into the corpus of the living law'.[8]

Modern legal systems have their origins in European civilization. Their 'formalism' is due to the increasing separation between morality and law since the eighteenth century and the dominance of positivism.[9] By the 1960s, when the first modern environmental laws were drafted, the separation was complete. Existing laws of unfettered resource use and exploitation – expressed in property rights, civil codes and public statutes – were complemented by conservation laws, but not in an integrated manner. They were not intended to alter unsustainable development.[10] The very existence of environmental law as a distinct subject area is proof of the fact that sustainability values have not found their way into the legal system. Both domestic and international environmental law are characterized by the absence of integration and ecological sustainability. Moreover, with their anthropocentric, resource-oriented and non-integrative approach, they tend to foster modern industrialism rather than changing it.[11]

The 'ecological ignorance'[12] of modern environmental law is widely recognized in the literature and needs no further description. What is important in our context, however, is the other tradition of European legal culture. The history of environmental law did not begin in the 1960s, but is as old as European legal history. There is a tradition of sustainability concepts that has, in fact, influenced the development of laws in European countries.

To this end, the history of sustainability is linked with the history of environmental law. Sustainability concepts have not been invented at the end of the twentieth century, but some 600 years earlier when continental Europe suffered a major ecological crisis.[13] Between 1300 and 1350 agricultural development and timber use had reached a peak that led to an almost complete deforestation.[14] The loss of ecological carrying-capacity had a number of severe consequences. Without forests there was no timber for heating, cooking, house building and tool making. At the same time, an important nutritional basis for deer, pigs and cattle vanished and with it the prospect for animal fertilizers necessary for growing crops. Erosion, flooding and lowering of water table levels were further effects. The resulting great starvations between 1309 and 1321, followed by the plague (the 'Black Death')

---

8   Ibid.

9   Bosselmann, K. (1995), *When Two Worlds Collide: Society and Ecology* (Auckland, RSVP), 227.

10   Ibid. 58–62, 80–87.

11   The fundamental critique of environmental legal theory has been developed in Bosselmann, K. (1985), 'Wendezeit im Umweltrecht', *Kritische Justiz* 345–61; Bosselmann, K. (1986), 'Eigene Rechte für die Natur?', *Kritische Justiz* 1–22; see further Bosselmann, K. (1992), *Im Namen der Natur* (Munich, Scherz).

12   Bosselmann 1995 (n. 9 above), 226.

13   Hughes, J.D. (2001), *An Environmental History of the World* (London, Routledge).

14   Abel, W. (1976), *Die Wüstungen des ausgehenden Mittelalters*, 3rd edn (Stuttgart, Fischer); Küster, H. (1998), *Geschichte des Waldes: Von der Urzeit bis zur Gegenwart* (Munich, Beck).

between 1348 and 1351, decimated the population of Middle Europe by one third.[15] In some regions half of the townships disappeared, altogether 40,000 settlements.[16] Compared to the collapse of other cultures, such as the Maya, the European collapse was less dramatic, but marked a significant 'cultural standstill'.[17]

In response to the crisis, local principalities and townships took measures of large-scale reforestation and enacted laws based on sustainability. The idea was to not clear more wood than would grow again and to plant new trees so that future generations would benefit. From the end of the fourteenth century, local laws in Middle Europe were guided by sustainability concerns.[18]

The approach to sustainability laws centred around a land use system known as 'Allmende' in German and 'commons' in English. Essentially, the land was seen as a public good setting limitations to individual land use rights. The *Allmende* system of German principalities defined the difference between the public and the private: the functioning and integrity of ecosystems was of public concern, the use of resources could be private. The rule was public ownership, the exception private use. This rule was eventually reversed in the nineteenth century when the model of private ownership became the rule. Ever since, public restrictions have only been possible in exceptional cases.

Under the *Allmende* system, land use rights were typically restricted in three ways. First, an important ecological limitation was the relational context of land use rights. They were regarded as heritage from the past and obligation for the future. Notions of heritage ('Erbschaft') for ancestors ('Ahnen'), on the one hand, and heirs ('Erben') and descendants ('Nachkommen'), on the other, commonly defined the extent of individual land use rights. Second, forests, pastures and arable land were organized as *Allmenden* in a narrow sense, i.e. as an undivided common area of the local community. Actual crop fields were allocated to individual households in terms of harvesting and possession, however, the decision on the kind of use remained with the local community. To this end, crop fields remained part of the *Allmenden*. It is important to note that local common ownership is different from private property as it can only be exercised through collective decision-making. Gary Hardin's famous 'tragedy of the commons'[19] is misleading, therefore, in so far as it refers to historical categories. Neither the German *Allmende* nor the English commons system allowed for excessive land use. It is more correct to speak of the 'tragedy of free access' typical for our time, i.e. since 1800.[20] The third restriction of land use rights came from the fact that they could not be sold or passed on without approval of the principal or local landlord (representing the collective). There was also the possibility of prohibiting changed or excessive use.

15 Herlihy, D. (1998), *Der Schwarze Tod und die Verwandlung Europas* (Berlin, Wagenbach).

16 Abel 1976 (n. 14 above).

17 Diamond 2005 (n. 2 above).

18 Marquardt, B. (2003), *Umwelt und Recht in Mitteleuropa: Von den großen Rodungen des Hochmittelalters bis ins 21.Jahrhundert* (Zürich, Schulthess).

19 Hardin, G. (1968), 'The Tragedy of the Commons', *Science* 162, 1243–8.

20 Sieferle, R.P. (1998), 'Wie tragisch war die Allmende?', *GAIA* 7:4, 304–7(4).

The concerns underpinning the *Allmende* system may have included property protection and political power. Importantly, however, they were also informed by ethics different from the ethics of modern property rights. The human–nature relationship was seen as one of stewardship. Land was respected as an essential ingredient of life in general with humans being mere users. Moreover, while land could be owned, it could only be owned within the limits of ecological sustainability.

Sustainability law of this kind was typically administered by the local principal and usually 12 judges forming the local court, but always within small communities. People were familiar with the social and natural environment they lived in. This had some major advantages. First, land use was highly decentralized and fully controlled by local communities. Second, intimate knowledge of local ecosystems allowed for informed decision-making. Third, form and extent of land use could easily be adjusted to changing ecological conditions. Fourth, common interests had preference over individual interests. And fifth, the aim of – often rotational – land use was not maximization, but optimization.

Judging by the success of sustainable forest and pasture management, environmental law had been rather effective until 1800. Around this time another major food and environmental crisis hit the now faster growing population in Europe. Simultaneously, traditional agricultural civilization gave way to modern industrial civilization. Only comparable to the Neolithic revolution, the Industrial Revolution led to a profound transformation of land and natural resource use.

This transformation had three different aspects:

1. The environmental aspect: the pressures of the demographic-ecological crisis caused the agricultural system to 'expand' its natural boundaries.
2. The philosophical aspect: Newton's model of physics coupled with the mechanistic-atomistic image of nature favoured 'natural resource' exploitation over ecological sustainability.
3. The energy aspect: renewable energy sources such as wood and wind were replaced by fossil energy, i.e. coal and later oil.

Together, these aspects became a catalyst for modern economy with its resource-intensive and short-term orientation. The law mirrored this shift. It left its sustainability-oriented, localized and public character more and more behind, adopting a 'private free enterprise' approach. At the beginning of the nineteenth century public environmental law virtually disappeared. The emerging system of private law and absolute property rights was largely ignorant of environmental protection, let alone sustainability. The Prussian Land Law of 1811, for example, granted individuals complete dispositional powers over their crops, land and ecosystems.[21]

The new property model subjected nature to exclusive private control. The relationship of humans to the land was no longer perceived as embedded in broader natural cycles, but as a relationship of individual power over the land. Now, the land user stood separate from, and above, nature. Consequently, any limitations of land

---

21   Marquardt, B. (2005), 'Zeitenwende für die Nachhaltigkeit: Zur umwelthistorischen Zäsur um 1800', *GAIA* 14:3, 243–52(10).

and property rights were no longer set by ecological necessities, but by competing rights of neighbours. It took governments until the second half of the twentieth century to realize that some environmental safeguards may, in fact, be needed. But even the new public environmental law of the 1960s and 1970s added only certain environmental duties to otherwise unrestricted private property rights. Until today, environmental law has remained the poor cousin of property and commercial law, only able to promote 'insufficient measures at the periphery'.[22]

The idea of sustainability may have vanished with the emergence of industrialism, but it has never died. Dormant until its global debut in the 1980s, the idea of sustainability has been alive for many hundreds of years in European civilization. Even the term 'sustainability' had been in use for several hundred years when the Brundtland Commission in 1987 employed it for its own definition of sustainable development.

The Term

We will now turn to the history of the term 'sustainability'. Even more clearly than the underlying idea, the term sustainability reveals its message for today: if you want long-term economic prosperity, look after the environment first!

The term sustainability was invented during the Age of Enlightenment. The enlightenment brought two important developments of thought:

1. The scientific revolution based on rational thinking and empirical observations. It revolutionized not only the way the physical world (nature) was perceived, but also the way the cultural world (society) was perceived. All aspects of human life could now be explained without regard to metaphysical levels of human existence (religion).
2. The secular approach to law and governance. Social norms (whether moral or legal) had to be 'reasoned', i.e. tested against rationality and scientific evidence. Tradition and religion ceased to be sources of public morality.

The dark side of such enlightened thought is the difficulty of reflecting on its own assumptions. If the rise of enlightened thought has brought unprecedented economic success, it has also brought unprecedented ecological failure. Finding ways out of this failure requires rethinking. It is doubtful that pure rationality provides sufficient guidance. The emergence of the sustainability concept in the eighteenth century, however, is a strong indication that modern rationality can be coupled with ancient wisdom.

The backdrop against which the term sustainability was coined was again the experience of an ecological crisis. Like the ecological crisis in the late Middle Ages, this new crisis was one of deforestation, this time caused by rapidly increased economic demands. Wood was needed for mining, shipyards, construction, manufacturing and household consumption. By 1650 widespread shortages of wood

---

22   Bosselmann 1995 (n. 9 above), 10.

began to cripple the economies in European countries.[23] At the same time, the new discipline of forest science and management emerged. Its focus was on studying the conditions for sustained forestry and sustainable yield.

Since 1662 the British Royal Society had worked for the Navy to investigate a sustained supply of timber for the development of a mighty maritime fleet. When, in 1666–67, England suffered a defeat of its naval fleet against the Netherlands, the Royal Society commissioned a report by one of its members that should summarize the debates within the Society. This member was the garden planner, biologist and historian John Evelyn. His report was published in 1664 under the title *Sylva, or a Discourse of Forest-Trees and the Propagation of Timber in His Majesties Dominions*. It made an instant impact and went to many subsequent editions.

In his analysis, Evelyn blamed the British glass and iron industry for excessive use of charcoal and the agricultural industry for 'disproportionate spreading of tilling' of forests to farmland. He saw the epidemic loss of forests as threatening to the 'magazines of timber' for the 'wooden walls' of the nation, i.e. the Royal Navy. He asked the gentry and landowners to show a 'new spirit of industry' by planting trees wherever possible: 'Let us arise then and plant!'[24]

The book details methods of planting, gardening, designing parks and managing forests. Citing the bible, classic philosophers like Plato and recent writers like Shakespeare, Evelyn argues the case for radical reforms including such ambitious ideas as shifting the entire iron industry from 'Old England' to 'New England', i.e. North America. The core of his argument is concern for 'posterity'. Each generation is 'non sibi soli natus' (not just born for itself), but 'born for posterity'.[25] In this context, Evelyn formulates his ethics of sustainability: 'men should perpetually be planting, so that posterity might have trees fit for there [sic] service ... which it is impossible they should have, if we thus continue to destroy our woods, without this provisional planting in their stead, and felling what we do cut down with great discretion, and regard to the future'.[26]

At the same time in France, the powerful Minister of Louis XIV, Colbert, promoted his 'grande reformation des forêts'. As in England, the French king sought to expand his naval fleet and was hit by the loss of timber through overuse by farmers and private households. Colbert's 'grande ordonnance' limited the rate of tree felling, restricted cattle farming and provided for tree planting programmes. The strategy was 'bon usage de la nature' as the essence of sustainable forest management.

German engineer and forest scientist Hans Carl von Carlowitz can be seen as the actual creator of the term sustainability – 'Nachhaltigkeit'. Born in 1645 in Freiberg, Saxony, he traversed Europe between 1665 and 1669 to gain insights into the resource crisis. He worked with John Evelyn in London where he also

---

23 Radkau, J. (2000), *Natur und Macht. Eine Weltgeschichte der Umwelt* (Munich, Beck), 245.

24 Quoted here from the 1776 edition of Evelyn, J. (1664), *Sylva, or a Discourse of Forest-Trees and the Propagation of Timber in His Majesties Dominions* (London, Jo. Martyn and Ja. Allestry), 279.

25 Ibid. 273.

26 Ibid. 205.

studied the British and French literature. For the following 40 years Carlowitz was responsible for administering the silver mining industry in Saxony. During this time he learned about the dependence of mining on its natural resource base. A year before his death (1714) he published a book that summarized his professional and lifetime experiences. The book's title is *Sylvicultura oeconomica oder Naturmässige Anweisung zur Wilden Baum-Zucht* [*Forest Economy or Guide to Tree Cultivation Conforming with Nature*]. Its subject is an investigation on 'how such conservation and growing of timber can be managed in order to provide continued, durable and sustained use'.[27]

This was the first appearance of the term sustainability/*Nachhaltigkeit*. The chapters preceding this appearance contain a critique of short-term gain-oriented forestry[28] and of blind cultivation of natural landscapes.[29] Carlowitz argues that ignorance and greed will 'ruin' forestry and lead to 'irreparable damage'.[30] He therefore demands sustainability as an 'indispensable thing' to ensure the continued 'existence of the country'.[31] Among his recommendations are 'the arts of saving wood'[32] through energy conservation, insulation, etc., systematic regrowing and planting of 'wild trees',[33] and the search for 'surrogates'[34] for wood such as peat and brown coal.

Fundamentally, Carlowitz builds his argument upon his perceptions of nature. In the first part of the book, he describes nature as 'mild'[35] and 'kind'[36] with a 'living spirit'[37] and the 'live instigating power of suns' to cause the 'miracle of vegetation'.[38] Nature can 'never be fully understood',[39] we can only observe 'how nature plays' and 'contemplate the amazing wonders of nature'.[40] Carlowitz rejects Descartes' view of nature as mere 'res extensa' and mere storage of resources. To him nature is alive, even 'inspired' like a conscious single organism. Such a holistic view does not hinder him from drawing his conclusions logically and rationally with a strong sense of practicality.

---

27   Carlowitz, H.C. von (1713), *Sylvicultura oeconomica. Anweisung zur wilden Baum-Zucht* (Leipzig, repr. Freiberg , TU Bergakademie Freiberg und Akademische Buchhandlung, 2000). See also Grober, U. (2002), 'Tiefe Wurzeln: Eine Kleine Begriffsgeschichte von "sustainable development" – Nachhaltigkeit', *Natur und Kultur* 3:1, 116–28; Grober, U. (2007), *Deep Roots – A Conceptual History of 'Sustainable Development' (Nachhaltigkeit)* (Berlin, Wissenschaftszentrum Berlin für Sozialforschung).

28   Carlowitz 2000 (n. 27 above), Foreword, and 43–98.
29   Ibid. Foreword and 79–94.
30   Ibid. 87.
31   Ibid. 106–7.
32   Ibid. 43–4 ('Holzsparkünste').
33   Ibid. 49.
34   Ibid. 425–30 ('Surrogate').
35   Ibid. 91.
36   Ibid. 113 ('gütig').
37   Ibid. 22 ('Lebens-Geist').
38   Ibid. 24.
39   Ibid. 31.
40   Ibid. 39.

In his economic perspective, Carlowitz accepts that humanity has lost paradise and cannot simply rely on nature delivering in abundance. Despite, or because of, the need of human intervention, the approach must be one of supporting nature and 'working with her',[41] not one of exploitation and 'acting against nature'.[42] The mistake to be avoided during economic progress is any wasteful, overusing and exploiting use of resources.

Carlowitz sees ecological conditions as determining for all human activities. In today's terminology, ecological integrity has to be respected and must not be compromised. Everything is to be measured against this rule. Carlowitz not only measures economic concerns against ecological sustainability, but also social concerns. His ethical beliefs are firmly grounded in social justice as part of ecological sustainability. The book's dedication expresses concern for 'poor subjects'[43] and 'dear posterity';[44] in fact, the entire book is a plea for responsibility towards future generations with many variations of the same theme. Remarkably, nowhere does Carlowitz assume any tension between social and intergenerational justice, on the one hand, and ecological sustainability on the other. His definition for long-term sustainability of the 'common sphere'[45] is to preserve the natural stock, which alone determines what humans can use now and in the future.

The *Sylvicultura oeconomica* considerably influenced forest management and cameralist theory in most German principalities. In 1757, the Württemberg cameralist Wilhelm Gottfried Moser published his two volumes of *Grundsätze der Forst-Oeconomie* (*Principles of Forest Economy*). Modifying Carlowitz's term 'nachhaltend' ('sustained') to 'nachhaltig' ('sustainable'),[46] Moser defines forest management on the basis of three principles:

1. 'sustainable economy with our forests';[47]
2. 'arts of wood saving';[48] and
3. 'regrowing planting and new wood growing'.[49]

He demands that these principles be applied with 'our children and posteriors in mind' adding that this is 'so reasonable, just, prudent and social, the more it is understood that no person lives only for themselves, but also for others and those to

---

41   Ibid. 31 ('mit ihr agiren').
42   Ibid. 39 ('wider die Natur handeln').
43   Ibid. Widmung ('armen Untertanen').
44   Ibid. ('lieben Posterität').
45   Ibid. ('des gemeinen Wesens').
46   An earlier use of this term can be traced back to the author of the 1729 *Forstlagerbuch*, von Göchhausen, who uses 'nachhaltig' ('sustainable') as the advanced form of 'pfleglich' ('appropriate' or 'considerate'); Schwarz, E., '*Oberlandjägermeister v Göchhausen*', *Archiv für Forstwesen* 9:7.
47   Moser, W.G. (1757), *Grundsätze der Forst-Ökonomie* (Frankfurt, Leipzig), 31. ('nachhaltige Wirtschaft mit unseren Wäldern').
48   Ibid. ('Holzsparkünste').
49   Ibid. ('Nachpflanzen und neuer Holz-Anbau').

come'.[50] Interestingly, the author is the son of Johann Jacob Moser (1701–85), one of Germany's most famous pioneers of civil rights and liberalism.

At the beginning of the nineteenth century sustainability was commonly accepted as synonymous with good forestry practice. The standard text on forestry by Heinrich Cotta (1763–1844), for example, refers to rich sustainability literature and mentions widespread sustainable forestry practice.[51] Georg Ludwig Hartig (1764–1837) describes the task of forest management as follows:

> es lässt sich keine dauerhafte Forstwirtschaft denken und erwarten, wenn die Holzabgabe aus den Wäldern nicht auf Nachhaltigkeit berechnet ist. Jede weise Forstdirection muss die Waldungen des Staates, ohne Zielverlust, taxieren lassen, und sie zwar so hoch als möglich, doch zu benutzen suchen, dass die Nachkommenschaft wenigstens ebenso viel Vortheil daraus ziehen kann, als sich die jetzt lebende Generation zueignet. [No long-term forestry can be thought of and expected if harvest of timber is not calculated with respect to sustainability. Every wise forest manager must appraise state forests without losing focus, that is at a highest possible yield level, but in a way that posterity benefits, at least as much as the living generation using it now.][52]

Cotta, Hartig and other classics of forest sciences had close affiliations with German idealism and holism (Leibniz, Schelling, Goethe, Herder, Hegel).[53] Cotta's first essay of 1792 outlines his organic worldview: 'In the whole world there is no thing without relationship to something else ... The world constitutes an indivisible whole; if we isolate something from the conditions in which it is embedded, we will disturb the order of nature and hinder its effects'.[54] Philosopher Johann Gottfried Herder (1744–1803) takes the Earth as the most suitable starting point for thinking about existence: 'Our Earth is a star among stars. Our philosophy of the history of the human species must begin from taking a view from the skies'.[55] Herder refers to the Earth as our 'Wohnplatz' ('living space' or 'home') and to humans' role in this house as 'haushalten' ('housekeeping'). This reference to the Greek term 'oikos' ('house', 'household') reflects European ancient wisdom and, at the same time, anticipates the terminology that is so familiar to us today. When Ernst Haeckel introduced the term 'Oecologie', in 1866, he used Herder's image of housekeeping: the Earth's living organisms relate to each other like inhabitants of a common house. 'Ecology' with its components 'oikos' and 'logos' describes the fundamental discipline of housekeeping against which 'economy' appears as a mere sub-discipline of efficient housekeeping.

Like Herder, Haeckel associated himself with both Darwin and Goethe. He saw no contrast between Darwin's mechanistic interpretation of evolution as survival of

---

50   Ibid.
51   Cotta, H. (1817), *Anweisung zum Waldbau*, 2nd edn (Dresden, Arnold). See Grober 2002 (n. 27 above), 123–4.
52   Hartig, G.L. (1795), *Anweisung zur Taxation der Forste oder zur Bestimmung des Holzertrags der Wälder* (Gießen), Foreword, quoted from Grober 2002 (n. 27 above), 124.
53   Bosselmann, K. (1992), *Im Namen der Natur* (Munich, Scherz), 12.
54   Grober 2002 (n. 27 above), 124.
55   Herder, J.G. (1784–1791), *Ideen zur Philosophie der Geschichte der Menschheit*, ed. Martin Bollacher (Frankfurt, Deutscher Klassiker, 1989), 21.

the fittest and Goethe's organic view of co-evolution as a collective process. Other natural scientists like Johann-Heinrich Jung-Stilling, Alexander von Humboldt and Georg Forster contributed to the eventual emergence of ecology as science.[56] They all shared the idea of sustainability as the appropriate way for humans to adapt to natural processes. Goethe influenced them all with his organic worldview. In a letter to Eckermann (11 April 1827) Goethe wrote: 'I think of the Earth as an enormous living being constantly breathing in and out'. One other metaphor captures the idea of sustainable economics: 'Baked bread is delicious and nurturing for one day, but flour cannot be sawn and seeds should not be used up through processing'.[57] Or, more bluntly, you may harvest the crop, but not the natural capital.

Sustaining life as a whole to provide for human life was a commonly shared view among nineteenth century theorists of forest management. This view was by no means confined to scholars and forest academies in Germany. Forest academies in Austria-Hungary, Switzerland, France, Russia, Scandinavia, the United Kingdom with its colonies and eventually the United States followed the same ecological concept. While the German word *Nachhaltigkeit* was understood abroad, it needed proper translation. The director of the French Forest Academy in Nancy, Professor Adolphe Parade, translated, in 1837, *Nachhaltigkeit* with 'production soutenu'. The French word 'soutenir' clearly shows its Latin roots. 'Sustinere' (from 'tenere' to keep) includes meanings such as endure, last, keep up, maintain, carry on, continue, sustain. The English word 'sustainable', too, captures these meanings and equals the French word.[58]

In his analysis of the terminological history, Ulrich Grober concludes that during the nineteenth century sustainability had emerged as the central term within forest sciences. It was now so broadly applied that the entire spectrum of the ecosystem 'forest' was included: location, fertility of soils, diversity of organisms, habitat for wildlife, water reservoir, protection against erosion, 'lung' function and recreational space. All these aspects were to be 'sustained' to capture the meaning of sustainability.

The sustainability principle was fundamental in forest legislation of the nineteenth century. Article 2 of the Bavarian Forest Act of 28 March 1852, for example, read: 'The management of state-owned forests has to follow sustainability as its highest principle'. The idea of sustainable management was not confined to forestry, but popularized among economists, planners and development theorists.[59] However,

---

56   Grober 2002 (n. 27 above), 125.

57   Goethe, J.W. von (1795), *Wilhelm Meisters Lehrjahre*, 'Lehrbrief' in Buch 7, Kapitel 9 ('Gebackenes Brot (ist) schmackhaft und sättigend für einen Tag; aber Mehl kann man nicht säen, und Saatfrüchte sollten nicht vermahlen werden'.)

58   Ironically, it took German authorities a long time to find a German equivalent for 'sustainable development'. Commonly used expressions include 'dauerhaft', 'zukunfsfähig' or 'tragfähig', before, in the late 1990s, the traditional term 'nachhaltig' (re-)emerged as the most widely accepted attribute of 'Entwicklung'.

59   Schanz, H. (1996),'Forstliche Nachhaltigkeit. Sozialwissenschaftliche Analyse der Begriffsinhalte und –funktionen', Ph.D. dissertation (University of Freiburg), 22–36; Eblinghaus, H. and Stickler, A. (1996), *Nachhaltigkeit und Macht. Zur Kritik von Sustainable Development* (Frankfurt, IKO), 43–4.

it was not sustainability that eventually prevented Germany from collapse. The pressures on forests were eased by sustainable forest management, but even more so by an economically driven shift to fossil energy. At the beginning of the nineteenth century wood and timber were largely being replaced by coal. Somewhat ironically, fossil fuel based industrialization came to the rescue of forests and prolonged the process of collapse.[60] From now on short-term cost–benefit thinking trumped sustainability.[61] As early as 1882 a forest expert could comment, therefore, that sustainability has been interpreted so broadly that it 'seems applicable to every form of exploitative economics'.[62]

Conclusion

Overlooking the tradition of sustainability in Europe, we can identify a consistent idea since medieval times. The idea is to live from the yield, but not from the substance. It has been closely associated with Europe's main resource base, the forests. Whenever this resource reached crisis point, governments looked for sustainability as the rescue strategy.

Not risking the 'substance' has always been the biggest challenge since the beginnings of agriculture and civilization. Virtually all agricultural societies went through ecological crises and eventually reached collapse.[63] The destruction of natural living conditions was not always the main reason. In fact, usually a variety of reasons contributed to crisis and collapse.[64] But central was always the lack of adapting to changing living conditions, in other words, the inability to – socially and economically – live within the boundaries of ecological systems.

Despite such consistent patterns, there is a fundamental difference between the agricultural and the modern industrial stages of civilization. The problem of ecological sustainability in the pre-industrial stage was very different from ours today. Agricultural societies were embedded in natural resource cycles and modified them for their purposes. The risk of losing the entire resource base did not normally occur and when it did, as in the case of Europe's forests, more radical management strategies were adopted. As shown above, in the period between the late Middle Age (1350) and the beginning of the modern age (1800) complex management strategies with strict norms and sanctions were used to protect the 'substance'. We can also say that the pre-industrial age had no other option. To ensure survival, unsustainable economics had to be detected quickly. Industrialism, by contrast,

---

60　Bachmann, G. (2006), 'Warum blieb der Kollaps im neuzeitlichen Deutschland aus?', *GAIA* 15:4, 260–63.

61　Grober 2002 (n. 27 above), 126.

62　Borggreve, B. (1888), *Die Forstabschätzung* (Berlin), 253.

63　Tainter, J.A. (1988), *The Collapse of Complex Societies* (Cambridge, Cambridge University Press); Tainter, J.A. (2000), 'Problem Solving – Complexity, History, Sustainability', *Population and Environment* 22:3, 3–41.

64　Alexander Dumandt counted no less than 210 reasons causing the collapse of the Roman Empire; Dumandt, A. (1988), *Der Fall Roms – Die Auflösung des römischen Reiches im Urteil der Nachwelt* (Munich, Beck), 695; see also Diamond 2005 (n. 2 above).

allowed rapid economic progress without paying the environmental costs – they could be externalized.

Today, we are in a profoundly different situation. The industrialized, globalized world has reached a level of complexity that makes quick fixes impossible. Using traditional sustainability methods will not make much difference. The higher complexity presents itself in environmental, social and economic terms. First, the resource crisis of today is global in its dimensions meaning that any local sustainability strategy is bound to fail if not followed through everywhere. Second, socio-economic relationships are no longer purely local. Everything we do in our local communities has effects on communities around the world, especially in poor countries. Third, the economy seems far removed, almost immune from its natural resource base. Where so much money is at stake, fertility of soils, diversity of life and stability of climate appear as a luxury we cannot afford.

As a result, sustainability has become a distant prospect, and this at a time when it is more needed than ever before in human history. In this situation, the fact that the Brundtland Report had such an impact on public debate is highly significant. Quite obviously, the world is now in search for a solution. No state and no corporate organization denies the importance of 'sustainable development', and the vast literature surrounding the subject would not exist, if sustainable development held no promise for saving us from collapse. Considering such basic consensus, it would be irresponsible to dismiss the concept of sustainable development purely because there is no agreement on what it means.

It is crucial, however, to realize the ecological core of the concept. Not realizing it means that social, economic and environmental interests have nowhere to go. There is only ecological sustainable development or no sustainable development at all. To perceive environmental, economic and social as equally important components of sustainable development is arguably the greatest misconception of sustainable development and the greatest obstacle to achieving social and economic justice.

If sustainable development really means nothing more than balancing between competing interests, we could be optimistic. To some extent, modern environmental law and policy have always been a balancing exercise between conflicting interests. The very subject 'environment' pertains the entire spectrum of human activities including direct ecological impacts, economic enterprises, social interactions and public policies. Depending on how narrowly or broadly the environment is legally defined, it will determine scope and integrative character of environmental law. Since its beginning in the 1960s, modern environmental law has broadened its subject from natural resources and conservation to environmental regimes and ecosystem management. Today, reconciliation between environmental values, property rights, social justice and commercial interests is in the centre of most environmental law cases. There is hardly a case or development that does not involve the breadth of environmental, social and economic concerns.[65]

It is true that the process towards integrated environmental law has been slow, patchy and incomplete. It is also true, however, that sustainable development has,

---

65   See Bosselmann, K. and Grinlinton, D. (eds) (2002), *Environmental Law for a Sustainable Society* (Auckland, New Zealand Centre for Environmental Law), vol. 1, vii-ix.

at best, accelerated, but not initiated this process. The novel aspect of sustainable development, in this regard, is that it seeks to integrate legal, political and institutional arrangements. How successfully this is being done may be debatable, but 'integration' is not a hallmark of sustainable development, it is a characteristic inherent to any development or management model dealing with environmental issues.

It will only be a matter of time for governments to have the necessary mechanisms for integrated decision-making in place. Taking the EU system of sustainable development governance as an example, we can clearly see a process of integrating environmental policies into other policy areas. This process includes constitutional, legal and institutional arrangements as well as a generally increased appreciation of sustainability concerns.[66] The same is true for a number of countries, where governments have introduced national strategies for sustainable development, intragovernmental offices, independent advisory boards, and environmental legislation containing the principle of sustainability.[67]

At international level, sustainable development is accepted as global policy and as 'an integrated part' of international environmental law.[68] And while there is no coherent body of 'sustainable development law', its 'principles, practices and prospects' can been described in support of the idea of integration.[69] It is perceivable, therefore, that the relationship between diverse areas such as environmental law, economic law and social law (including human rights) can be clarified and strengthened[70] to eventually create some form of integrated law.

The big question is, however, whether integration of laws will be sufficient to achieve what sustainable development demands. Using an analogy, is the outcome of a custody case determined by considering all the various personal, social and economic issues that may be involved? Or is it not the 'best interest of the child' that ultimately determines the outcome? Of course, the 'best interest' is the ultimate

66   Bosselmann, K. (2003), 'The Environmental Governance of the European Union: Institutional and Procedural Aspects of Sustainability', in Lilly, I. and Bosselmann, K. (eds), *Repositioning Europe: Perspectives from New Zealand* (Christchurch, NCRE), vol. 2, 1–19.

67   Bosselmann, K. (2007), 'Why New Zealand Needs a National Sustainable Development Strategy', in Parliamentary Commissioner for the Environment, *Sustainability Review 2007: New Zealand's Progress Toward Sustainable Development*, Background Paper, Wellington, 33; Weidner, H., Jänicke, M. and Jorgens H. (eds) (2002), *Capacity Building in National Environmental Policy: A Comparative Study of 17 Countries* (Heidelberg, Springer Verlag), available at <http://www.pce.govt.nz/projects/susstrategy.pdf>, accessed 1 December 2007.

68   Sands, P. (2003), *Principles of International Environmental Law*, 2nd edn. (Cambridge, Cambridge University Press).

69   Cordonier Segger, M.C. and Khalfan, A. (2004), *Sustainable Development Law: Principles, Practices and Prospects* (Oxford, Oxford University Press).

70   See Principle 4 of the Rio Declaration on Environment and Development (the Rio Declaration) (Rio de Janeiro, 13 June 1992; UN Doc. A/Conf.151/26 (vol. I); 31 ILM 874 (1992)); Chapter 39 of Agenda 21: Programme of Action for Sustainable Development (Agenda 21) (Rio de Janeiro, 14 June 1992; UN Doc A/Conf.151/26 (1992) 31 ILM 874 (1992)); Chapter XI of the Johannesburg Plan of Implementation (Johannesburg, 4 September 2002, UN Doc. A/Conf.199/20).

benchmark; all other factors are relevant and important, but subject to serving the child's best interest.

Likewise, if sustainable development would be used merely for integrating and balancing conflicting interests, nothing would be achieved. Without a benchmark, we are left at a guess *how* environmental, social and economic interests should be balanced. The Brundtland definition offered some direction by demanding not to compromise 'the ability of future generations to meet their own needs'.[71] This leaves, however, the question of needs and how they could be perceived for future generations. As will be shown now, the answer to this question is critical to both the idea of sustainability and the concept of sustainable development.

The Development Since 1972

In this sub-chapter we will take a closer look at the concept of sustainability as distinguished from the concept of sustainable development. Given the terms' origins as an ecological principle, the Brundtland Commission could have first defined 'sustainability' and *then* use it for its description of sustainable development. But this did not happen. Instead, the Brundtland Commission chose to introduce a quite different, broader concept. Both advocates and critics of the concept of sustainable development agree that the original meaning of sustainability has been obscured by the Brundtland definition. They disagree, however, on whether or not sustainability has been replaced by sustainable development so that it is no longer relevant for policy and law-making.

Early Documents Reflecting Sustainability

The modern history of sustainability is closely associated with the history of international environmental policy and law. The year 1972 marked their starting point. In that year, the Club of Rome published its report *The Limits of Growth*,[72] the UN Conference on the Human Environment took place in Stockholm and the United Nations Environment Programme (UNEP) was established in Nairobi. The Club of Rome saw economic growth on a conflict course with ecological sustainability; the UN system, on the other hand, believed in reconciling the two. International environmental law has followed this later assumption which is not surprising given the traditional nexus between growth, states and international law. Fundamentally, however, the environmental challenge required international cooperation of all, states, business and civil society. The environmental movement clearly had an impact on the shaping of key legal principles including precaution and sustainability. In can be said, therefore, that international environmental law has emerged as a new legal field created by science, philosophy, ethics, economics and politics.[73] Yet, despite its multidisciplinary

---

71   Brundtland Report (n. 4 above), 43.

72   Meadows, D.H., Meadows, D.L., Randers, J. and Behrens, W. III (1972), *The Limits to Growth* (New York, Universe Books).

73   See, for example, Kiss, A. and Shelton, D. (2000), *International Environmental Law*, 2nd edn (New York, Transnational Publishers), 11–25; and Sands 2003 (n. 68 above), 3–17.

components and new principles, international environmental law remained a mere branch of public international law. As such it was never in the position to break up the systemic linkages between economic growth, states and international law. States have continued to promote the compatibility of growth with sustainability.

The critique against this ideology or paradigm was at the centre of the growth debate that began in the 1970s. The 'limits to growth' paradigm elevated ecologism to a political theory challenging capitalism and socialism alike. From now on the divide between growth critics and growth followers became visible and continued to shape the sustainability debate of the 1980s and 1990s.

The continuity between the growth debate and the subsequent sustainability debate is often overlooked, but critically important for our context.[74] If a worldwide awareness of environmental issues emerged in the 1970s, there was also the emergence of two directions of environmentalism that never reconciled their positions. On the one hand, the critique of the growth paradigm has inspired those who envisaged sustainability as the counter-model to economic dominance. On the other hand, connecting the 'environment' (= sustainability) with 'development' (= growth) found friends in all political camps. The promise here was that environmental protection will only be successful if it goes hand in hand with economic prosperity. Growth critics such as Edward Goldsmith,[75] Mihajlo Mesarovic, Eduard Pestel,[76] Dennis Meadows,[77] Rudolf Bahro[78] or Herman Daly[79] have always been opposed to sustainable development, while authors like Wilfred Beckerman,[80] K. Arrow,[81] Peter

---

74   See Steurer, R. (2001), 'Paradigmen der Nachhaltigkeit', *Zeitschrift für Umweltrecht und–politik* 4, 537–39.

75   Goldsmith, E., Allen, R. Allaby, M., Davoll, J. and Lawrence, S. (1972), *Blueprint for Survival* (London, Tom Stacey); Goldsmith, E. (1992), *The Way: An Ecological Worldview* (London, Rider).

76   Mesarovic, M. and Pestel, E. (1972), *Menschheit am Wendepunkt* (Stuttgart, Deutsche Verlagsanstalt); Mesarovic, M. and Pestel, E. (1983), 'Organisches und dauerhaftes Wachstum', in Peccei, A., Pestel, E. and Mesarovic, M. et al. (eds), *Der Weg ins 21 Jahrhundert. Alternative Strategien für die Industriegesellschaaft, Berichte an den Club of Rome* (Munich, Molden), 81–97.

77   Meadows, Meadows, Randers and Behrens III 1972 (n. 72 above); Meadows, D.H., Meadows, D.L. and Randers, J. (1992), *Beyond the Limits: Confronting Global Collapse, Envisioning a Sustainable Future* (Vermont, Chelsea Green Publishing).

78   Bahro, R. (1972), *Die Alternative* (Frankfurt, European Publishing House Cologne); Bahro, R. (1994), *Avoiding Social and Ecological Disaster: The Politics of World Transformation: An Inquiry into the Foundations of Spiritual and Ecological Politics*, tr. David Clarke (Bath, Gateway).

79   Daly, H. (1991), 'Elements of Environmental Macroeconomics', in Constanza, D. (ed.), *Ecological Economics: The Science and Management of Sustainability* (New York, Columbia University Press), 32–46.

80   Beckerman, W. (1974), *In Defence of Economic Growth* (London, Jonathan Cape); Beckerman, W. (1995), *Small is Stupid: Blowing the Whistle on the Greens* (London, Duckworth).

81   Arrow, K., Bolin, R., Costanza, P. et al. (1995), 'Economic Growth, Carrying Capacity, and the Environment', *Science* 268, 520–21.

Bartelmus,[82] David Pearce[83] or William Nordhaus[84] saw growth as an inherent part of the new concept of sustainable development.

For the sake of simplicity, we can distinguish between the ecologist approach and the environmental approach to sustainable development. The ecologist approach is critical of growth and favours ecological sustainability (*'strong sustainability'*). The environmental approach assumes the validity of growth and places equal importance on environmental sustainability, social justice and economic prosperity (*'weak sustainability'*). The difference between them is not just gradual, but fundamental, as becomes clear when we follow the sustainability debate through to the 1980s and 1990s.

The term sustainability had not been used when international law took its first steps to integrate environment and development. Principle 13 of the 1972 Stockholm Declaration[85] called on states to adopt 'an integrated and co-ordinated approach to their development planning as to ensure that their development is compatible with the need to protect and improve the human environment'. Importantly, the essence of sustainability is also being referred to. Principles 3 and 5 called for the non-exhaustion of renewable natural resources and the maintenance and improvement of 'the capacity of the earth to produce vital renewable resources'.

A number of subsequent regional treaties reflected the same idea of integration, however, without implying a sustainability approach. The 1974 Paris Convention, for example, calls for an 'integrated planning policy consistent with the requirement of environmental protection',[86] the 1978 Kuwait Convention calls for an 'integrated management approach'[87] to achieve 'environmental and developmental goals in a harmonious manner' and the 1985 ASEAN Convention calls for ensuring that 'conservation and management of natural resources are treated as an integral part of development planning'.[88]

All these documents promote the idea of integrating policies rather than integrating the subject matters that policies are concerned with. This is an important difference. Policy integration is possible merely through law, administration and governance. The integration of subject matters, on the other hand, is a lot more complex. What is involved when the natural 'environment' and human 'development' are to be integrated? There are ethical choices to be made, for example, treating the natural environment as a basis and limitation for human development or human development as a basis and limitation for the natural environment. Significantly, the 1978 Amazonian Treaty makes such a choice by calling for a 'balance between

---

82   Bartelmus, P. (1994), *Environment, Growth and Development: The Concepts and Strategies of Sustainability* (Florence, KY, Routledge).

83   Pearce, D., Markandya, A. and Barbier, E.B. (1989), *Blueprint for a Green Economy* (London, Earthscan).

84   Nordhaus, W. (1993), 'Economic Growth on a Planet under Siege', in Siebert, H. (ed.), *Economic Growth in the World Economy* (Tübingen, Mohr), 223–42.

85   Declaration of the United Nations Conference on the Human Environment (Stockholm Declaration) (Stockholm, 16 June 1972; UN Doc. A/Conf.48/14, 11 ILM 1461 (1972)).

86   Paris Convention 1974, Article 6(2)(d).

87   Kuwait Convention 1978, Preamble.

88   ASEAN Convention 1985, Article 2(1).

economic growth and conservation of the environment'.[89] This goal-oriented approach to integration reveals the subject matter, that is, finding a balance between conflicting goals.

To include environmental and developmental goals in a description for integrated policies is a significant step forward. A further step is to find a single term for goal-oriented policy integration. Terms that roll environment and development into one include 'environmentally sound development' or 'ecodevelopment', and both have been used in the early 1980s. However, the most popular term became 'sustainable development'. As shown above, the term sustainability had been well established referring to the preservation of the substance or integrity of ecological systems. In this sense, development needs to be within the boundaries of ecological systems to qualify as 'sustainable'.

This meaning was also behind 'sustainable development' when it first appeared in an official document. Paragraph 2 of the 1980 *World Conservation Strategy* (WCS)[90] defines 'development' as 'the modification of the biosphere and the application of human, financial, living and non-living resources to satisfy human needs and improve the quality of human life' and adds: 'Among the prerequisites for sustainable development is the conservation of the living resources'. To this end, sustainability is preserved in its original meaning. However, the WCS then proceeds to set further requirements: 'For development to be sustainable it must take account of social and ecological factors, as well as economic ones; of the living and non-living resource base; and of the long-term as well as the short-term advantages and disadvantages of alternative actions'.[91] The aim of this definition was to introduce a new model of development, one that is not inherently threatening the 'living and non-living resource base'. The linking of social and economic factors to the notion of 'sustainable' could be seen as confusing; however, the overall concern of the WCS clearly is for ecologically sustainable development.

This concern was also behind the World Charter for Nature adopted by the UN General Assembly in 1983.[92] The Charter followed on from the 1972 Stockholm Declaration, but expressed the ethical approach more clearly. In setting out the principles 'by which human conduct affecting nature is to be guided and judged',[93] the Charter defined nature conservation as a prerequisite for the use of natural resources and development planning. Notably, it described humanity as 'part of nature' and states that 'every form of life is unique, warranting respect regardless of its worth to man'.[94] This approach leads to the requirement that natural resources may be managed to 'achieve and maintain optimum sustainable productivity', but must not be utilized

---

89   Amazonian Treaty 1978, Preamble.

90   International Union for Conservation of Nature and Natural Resources (IUCN) (ed.) (1980), *World Conservation Strategy: Living Resource Conservation for Sustainable Development* (Switzerland, IUCN, Morges).

91   Ibid. para. 3, 2.

92   World Charter for Nature (GA Res. 37/7, UN GAOR, 37th session, 48th plenary meeting, UN Doc.A/37/7 (1983)).

93   Ibid. Preamble.

94   Ibid.

'in excess of their capacity for regeneration'.[95] Like the Stockholm Declaration, the World Charter for Nature was not binding international law, but represented a soft law consensus among states. A treaty for sustainable development, if negotiated in the early 1980s, would probably have followed a definition for development based on ecological sustainability.

The Brundtland Approach to Sustainability

The UN World Commission on Environment and Development (WCED) took a quite different approach when, in 1984, it started its work towards the report *Our Common Future*. With members from 21 'very different nations',[96] many of them from the 'South', the WCED was not mainly concerned with ecological sustainability, but with two different set of problems. One set centred around global environmental degradation, the other around the discrepancies of social and economic development between 'North' and 'South'. Obviously, both problem areas are linked by development issues, but the challenge was to reconcile the North's unsustainable 'over-development' with the South's 'under-development'. This required a formula for development reflective of both ecological sustainability and social-economic justice.

The WCED tried to meet this challenge by pursuing two different agendas that eventually had to be reconciled. Large parts of the Brundtland Report are informed by the search for ecological sustainability. The Report criticized the 'accelerating pace and expanding scale of impacts on the ecological basis of development'[97] and went on to explain how wealthy countries' 'forms of development erode the environmental resources upon which they must be based' and how environmental degradation undermines economic development.[98] Calling for a realignment of humanity's relationship with the environment, the Report demands a 'new ethic to guide state behaviour in the transition to sustainable development'.[99] The Report also notes that 'it is futile to deal with environmental problems without a broader perspective that encompasses the factors underlying world poverty and international inequality'.[100] The 'common interest' in environmentally sound development cannot be promoted if there is a 'neglect of economic and social justice within and amongst nations'.

Essentially, the Brundtland Report is a plea for comprehensive distributive justice between (a) rich and poor, (b) people living today and in the future, and (c) humans and nature. This political plea is summarized in the famous phrase: 'Sustainable development is development that meets the needs of the present without compromising the ability of future generations to meet their own needs'.[101] Whether

---

95 Ibid. paras 4 and 10(a).
96 Brundtland Report (n. 4 above), xi.
97 Ibid. 3.
98 Ibid.
99 Ibid. 332.
100 Ibid. 43.
101 Ibid.

such general description provides sufficient guidance is debatable and has been questioned until today.

For the meaning of the principle of sustainability, this description neither adds nor detracts anything. It is simply silent about its meaning and central importance. Given the principle's ecological core, it is problematic, therefore, to think of sustainable development purely in terms of human needs. Such focus has been criticized as being overly anthropocentric. Anthropocentric limitations have certainly shaped the 1992 Rio Declaration.[102] However, the Brundtland Report is hardly responsible for such reductionism. It explains its own description in the following way: 'It contains within it two key concepts: the concept of "needs", in particular the essential needs of the world's poor … and the idea of limitations imposed by the state of technology and social organization on the environment's ability to meet present and future needs'.[103] The first key concept is the recognition that development should meet basic human needs, in particular, of the poor. This can be seen as the social aspect of sustainable development. The second key concept is the recognition that human activities (state of technology and social organization) must not ignore environmental limitations. This can be seen as the ecological aspect of sustainable development. Quite obviously, sustainable development has strong human connotations, but human needs can only ever be met within ecological boundaries. This seems the forgotten message of Brundtland.

The two concepts are related, but how? Are they equally important or is one more important than the other?

Sustainability Versus Development?

A widely held view is that both concepts are equally important. The assumption here is that human needs can only be met if environmental and developmental goals are pursued together: how could a protected environment meet the basic needs of the poor, if no development takes place, and how can development be beneficial, if it comes at the price of losing the environment? Structurally, this argument reflects a two-scales model of sustainable development. The environment sits on one side of the scales, development on the other; the art is to keep both in balance.[104]

There are three major problems with the two-scales model. First, it assumes a separation between the environmental and developmental spheres that does not exist in reality. Development is not a static entity, nor is the environment. The actual aim of sustainable development, to bring both spheres together, cannot be reflected in a model that aims for the balancing of two separate entities. Second, the time dimension, so essential to sustainability, is missing in a model concerned with the balancing of present issues at hand. What if both the environment and development

---

102   Ibid., for example, Principle 1: 'Humans are in the centre of concerns for sustainable development'.

103   Ibid.

104   The idea of keeping the balance is also behind the three-pillars model of sustainable development which is why the criticism against the two-scales model equally applies to the three-pillars model. See below.

are presently in a bad shape? Third, the equal-importance theory is ideologically biased. It reflects the liberal, and neoliberal, equation of development with economic growth and prosperity. This equation is not necessarily relevant to all people living now, for example, in the 'South', or in the future.

In reflection to the Brundtland Report, these three problems can be summarized as overlooking the WCED's main concern, that is, the long-term perspective of humanity on this planet. This makes it impossible to treat the 'two key concepts' of development and environment independently from each other. If this were the case, then human needs (of today and in the future) could be met either by Western-style economic development on a global scale – regardless of its environmental impacts – or by a total stop of present development to allow quick recovery of environmental systems. Both these extremes could serve human needs, perhaps even in the future, but only in a very limited understanding of what these needs may be. The WCED explained that 'the key element of sustainable development is the recognition that economic and environmental goals are inextricably linked'.[105] This prohibits any self-isolation of economic development from the environment. Obviously, the aspect of 'needs' must be seen with respect to economic as well as environmental preconditions.

The above-mentioned first key concept of the Brundtland description refers to the development problem ('needs'), while the second key concept refers to the sustainability problem ('environment's ability'). How then can 'needs' and the 'environment's ability' be defined?

The Question of Needs

There is no obvious reason to assume that only basic material needs are meant (healthy living conditions or protection from poverty). With equal justification one could include immaterial needs such as freedom, security, education or justice. And what about protection from the ever-increasing debts that the countries are presently producing? All these and many other things that we do today to meet our 'needs' have the effect of 'compromising the ability of future generations to meet their own needs'. While intra- and intergenerational equity are undoubtedly part of sustainable development, we are still in need of a clear benchmark against which these equity concerns can be measured. Without it, the implementation of sustainable development becomes an illusion and will not be noticeable.

Searching for a benchmark, the 'environment's ability' is the only credible choice. If the economy–environment relationship is the key element of sustainable development, as the WCED claims, and if the environment's ability cannot be compromised (to meet present and future needs), then the economy ('the state of technology and social organizations') must ensure the environment's ability. This excludes compromises or trade-offs between economic goals and environmental goals. In other words, only environmentally sound development could possibly meet present and future needs. The principle of sustainability is, therefore, paramount.

---

105   Brundtland Report (n. 4 above).

The same conclusion can be reached by concentrating on 'needs', which are so central to the Brundtland *definition*. It may be easy to determine the needs of people living today. Everyone needs access to water, food and shelter – essential needs that the world's poor are often not able to meet. It is equally safe to assume that people living in the future have the same needs. Beyond such essentials, however, it is difficult, if not impossible, to make assumptions about the needs of future generations. They are likely to differ from our needs today; however, we can only speculate in what way and to what extent. Rather than speculating about what fundamentally is an unpredictable future, we should concentrate on what we do know. We do know that humans have been, in the past, and will be, in the future, dependent on favourable living conditions. Without basic environmental services no human life is possible. Again, we cannot be too specific about the kind of environmental services that may be needed in the future, but accessible water, clean air, fertile soils and biological diversity are certain to be included here. These services, and an incalculable number of additional services, constitute the 'environment's ability' to meet human needs today and in the future.

At this point, it should become clear how severely hampered anthropocentric approaches to sustainable development are. They attempt the impossible. How could the needs concept ever be practical if based on pure speculation about the future? Human-centred motivations for a planet that *we* may need in the future are riddled with uncertainties. The unqualified limitation to *human* needs leaves speculations wide open ranging from a future without nature right through to a future without humans. It is not beyond imagination that humans will, one day, have completely substituted their natural environment with an artificial environment. And how could a desire of the human species be excluded to vanish from the face of the Earth? Whether the former or the latter is favoured may be a matter of (bad) ethical choice, but both variants are expressions of anthropocentric reductionism. They overlook a simple truth, namely the interrelatedness of all life across the boundaries between humans and non-humans.

The concern for all forms of life, not just human life, therefore, is a better guide into the future. It allows for a more practical focus on essentials common to all life, i.e. the ability to exist, reproduce and evolve. These essentials call for the preservation of conditions that have proven to be favourable in the past. And while the prediction of the relevance of such conditions for future life is far from reliable, it is a lot easier to make than trying to predict what future human generations might see as their needs.

It makes good sense to think of the future of life and not just human life. From an ecological point of view, anthropocentric limitations are difficult to justify. Edith Brown Weiss, the eminent proponent of intergenerational equity,[106] has never attempted to exclude the welfare of nature from our thinking about future generations. Her three principles of intergenerational equity involve the preservation of options, quality and

---

106   Brown Weiss, E. (1989), *In Fairness to Future Generations: International Law, Common Patrimony and Intergenerational Equity* (Tokyo, United Nations University Press), 17 et seq.

access.[107] The first principle requires 'to conserve the diversity of the natural and cultural resource base', the second 'to maintain the quality of the planet' and the third to provide 'equitable rights of access to the legacy of the past'.[108] While this third principle is anthropocentric in character, the former two describe duties towards the community of life. There is certainly no suggestion in Brown Weiss's concept to be only concerned with the welfare of humans.[109] According to Ulrich Beyerlin 'much speaks in favour of conceiving the intergenerational component of sustainable development in eco-centric terms'.[110] He concludes: 'As intergenerational equity is inseparably intertwined with intra-generational equity, the concept of sustainable development in its entirety must be perceived as both anthropocentric and eco-centric in nature'.[111]

Other commentators have made the same point, thus challenging the widely held view that sustainable development reflects anthropocentrism and trade-offs between environmental, economic and social interests.[112] The fundamental importance of the environment is not contingent on an ecological approach to sustainable development. The economic approach may emphasize growth and prosperity, but can nevertheless be formulated on the basis of ecological sustainability. World Bank economist Roberto Repetto, in his 1986 book *World Enough and Time*, wrote 'the core idea of sustainability is that current decisions should not impair the prospects for maintaining or improving future living standards. This implies that our economic systems should be managed so that we can live off the dividends of our resources'.[113] Two other economists from the World Bank, Mohan Munasinghe and Ernst Lutz, define

---

107   Brown Weiss, E. (1990), 'Intergenerational Justice and International Law', in Busuttil, S. et al. (eds), *Our Responsibilities to Future Generations* (Malta, Foundation for International Studies), 98 et seq.

108   Ibid.

109   Bosselmann, K. (2002), 'A Legal Framework for Sustainable Development', in Bosselmann, K. and Grinlinton, D. (eds) (2002), *Environmental Law for a Sustainable Society* (Auckland, New Zealand Centre for Environmental Law), vol. 1, vii-ix, 151.

110   Beyerlin, U. (2006), 'Bridging the North-South-Divide in International Environmental Law', *Zeitschrift für ausländisches und öffentliches Recht* 66, 263–75. See also Agius, E. (1998), 'Towards a Relational Theory of Intergenerational Ethics', in Busuttil, S. et al. (eds), *Our Responsibilities to Future Generations* (Malta, Foundation for International Studies), 87; and Taylor, P. (1998), *An Ecological Approach to International Law* (London and New York, Routledge), 281–2.

111   Beyerlin, U. (2006) (n. 110 above).

112   See Boer, B. (1995), 'Implementation of International Sustainability at a National Level', in Ginther, K. et al. (eds), *Sustainable Development and Good Governance* (Dordrecht and Boston, Kluwer Academic Publishers), 111–13; Taylor 1998 (n. 110 above), 325–7, 348–9; Gillespie, A. (1997), *International Law, Policy and Ethics* (Oxford, Oxford University Press), 2–5, 127–8; and (with less clarity) Birnie, P. and Boyle, A. (2002), *International law and the Environment*, 2nd edn (Oxford, Oxford University Press), 44–7; Susan Murcott's list of 57 different definitions of sustainable development (from 1979 to 1997) contains some 20 definitions that emphasize ecological sustainability as the basis of any form of development, available at <www.aocweb.org/emr/Portals/2/MIT%20Definitions.pdf>, accessed 1 December 2007. See also contributions in: Bugge, H.Ch., and Voight, Ch. (eds) (2008) *Sustainable Development in International and National Law* (Groninger: Europa law Publishing).

113   Repetto, R. (1986), *World Enough and Time* (New Haven, Yale University Press).

'sustainable development as an approach that will permit continuing improvements in the quality of life with a lower intensity of resource use, thereby leaving behind for future generations an undiminished or even enhanced stock of natural resources and other assets'.[114] Clearly, the preservation of the natural stock determines the ability to meet the needs of present and future generations.

The eco-centric component of sustainable development is indeed crucial for making the concept operable. If we perceive human needs without regard to ecological reality, we are at risk of losing the ground under our feet. Against this reality, any talk about the equal importance of development and environment, the two-scale model, 'three-pillar model' or 'magic triangle', is pure ideology. The concerns for social justice and economic prosperity are valid and important, but secondary compared to the functioning of the Earth's ecological systems. Ecological sustainability is a prerequisite for development and not a mere aspect of it.

Therefore, despite the ambiguity of the Brundtland Report and its reception in the literature and international documents, it would be wrong to conclude that the principle of sustainability has lost its contours. Rather, we need to analyse very carefully how the principle has been used in the various declarations and principles of international law. Its continued presence would give us the evidence for the thesis that sustainability is a fundamental principle of law.

A few years after the Brundtland Report, the World Conservation Strategy was revised. The 1991 document *Caring for the Earth: A Strategy for Sustainable Living*[115] incorporated the Brundtland Report, but did not lose track of the core meaning of sustainable development. It describes the purpose of sustainable development as improving the quality of human life while living within the carrying capacity of the Earth's ecosystems.[116] *Caring for the Earth* contains two distinct requirements: one is the commitment to a new ethic based on respect and care for one another and for the Earth, the other is integration of conservation and development.[117] The document warns not to lose the focus of sustainability when it is used in combination with other words such as 'use', 'yield', 'development', 'management', 'economy' and 'society'.[118]

The Declarations from Rio and Johannesburg

Until 1992, there was no apparent rift between *strong* and *weak sustainability*. Wherever the term sustainability was used, it had the meaning of ecological sustainability, and where the term sustainable development was used, the principle of sustainability was implied. A rift emerged, however, in June 1992 during the Rio UN Conference on Environment and Development (UNCED). Symbolically visible

---

114   Munansinghe, M. and Lutz, E. (1991), *Environmental-Economic Evaluation of Projects and Policies for Sustainable Development*, Environment Working Paper No. 42 ([location], The World Bank).

115   International Union for the Conservation of Nature and Natural Resources (IUCN), World Wide Fund for Nature (WWF) and United Nations Environment Programme (UNEP) (1991), *Caring for the Earth: A Strategy for Sustainable Living* (Gland, Switzerland, IUCN).

116   Ibid. 3–4.

117   Boer 1995 (n. 112 above), 111, 113.

118   IUCN, WWF and UNEP (n. 115 above).

in the fortress-type venue of the official Earth Summit, where states negotiated, and the beachside venue of the Global Forum, where civil society groups met, the sustainability agendas differed considerably. The 160 participating states left any commitment to sustainable development to two soft law documents, the Rio Declaration on Environment and Development[119] and Agenda 21,[120] both stressing the interconnectedness of environmental, social and economic concerns. Civil society, on the other hand, emphasized ecological sustainability as the key to dealing with social and economic concerns. Several hundred NGOs negotiated 15 so-called 'alternative treaties', among them the Earth Charter, as an alternative to the Rio Declaration. Ecological sustainability was referred to as central to everything: poverty eradication, socio-economic development, human rights and peace.

The Rio Earth Charter was not only responding to the Rio Declaration, but to the entire preparatory process for UNCED. Following the Brundtland Report with its call for a 'new charter' that 'should prescribe new norms for state and state behaviour to maintain livelihoods and life on our shared planet',[121] several preparatory meetings in 1990 and 1991 identified the elements for an Earth Charter to be agreed upon by states. However, a few months before the Rio Summit, at the fourth and last Preparatory Committee (PrepCom), it became clear that intergovernmental agreement could not be reached. States moved away from the idea of a Charter with its emphasis on ecological sustainability and eventually adopted the Rio Declaration[122] instead.[123]

This is not say, however, that the concept of sustainability was dismissed. The Rio Declaration just lacked definitions for sustainable development or sustainability, instead favouring a comprehensive description of political issues involved. A careful interpretation shows that the Rio Declaration did not promote the three-pillars model that was heralded by many post-Rio commentators. Principle 4 of the Declaration states 'that in order to achieve sustainable development, environmental protection shall constitute an integral part of the development process and cannot be considered in isolation from it'. Thus, the integration of environmental and developmental issues is the *modus operandi* of sustainable development. However, Principle 4 also conceptually relates development to the environment: it cannot be pursued independently of preserving the natural resource basis.

The Declaration mentions the right to development (Principle 3),[124] the integrative approach (Principle 4) and the indispensable task of poverty alleviation (Principle 5)

---

119   Taylor 1998 (n. 110 above), 324–7, 379–81. The Rio Earth Charter was followed up by the 1994 initiative of Maurice Strong, Chair of the Earth Council, and Mikhail Gorbachev, President of Green Cross International. The new initiative led to a worldwide process of consultation with the eventual launch of the Earth Charter in 2000. See Vilela, M. (2005), 'Building Consensus on Shared Values. History and Provenance of the Earth Charter', in Blaze Corcoran, P. (ed.), *The Earth Charter in Action. Toward a Sustainable World* (Amsterdam, KIT), 17 et seq.

120   Agenda 21 (n. 70 above).

121   Brundtland Report (n. 4 above), 332.

122   Rio Declaration (n. 70 above).

123   Vilela 2005 (n. 119 above), 18.

124   The proximity to the Brundlandt definition with its 'needs' concept is obvious in the wording: 'The right to development must be fulfilled so as to equitably meet developmental

as key aspects of sustainable development. The relative prominence of the right to development and poverty alleviation are clearly meant to acknowledge the special needs of developing countries. However, this does not diminish the importance of ecological sustainability. The Declaration does not allow industrialized states to compromise the principle of sustainability for the sake of pursuing their economic prosperity and social development.[125] The developing states participating in Rio did not insist on their right to development to allow multinational corporations of rich states to develop more quickly and effectively. The political implication merely is to not expect sustainability measures from developing states with the same urgency as can be expected from developed states that have already affected the Earth's ecological integrity to a much higher degree.

Essentially, the Rio Declaration attempted to find a solution to the distribution problem that the global ecological crisis has created. This attempt may have been incomplete and not very satisfactory; however, it can hardly be argued that the principle of sustainability had been dismissed or replaced by something else.[126] To the contrary, Principles 3 and 4 confirm its existence: even economic development in developing countries must not compromise the ability of future generations to meet their needs.[127] If the right to development could only be exercised by violating this principle of sustainability because of developed countries' overburdening of the environment, then *these* countries need to reduce their overburdening in order to preserve the development chances of developing countries. It follows that ecological sustainability is paramount in the developing world, and even more so in the developed world.

The other Rio soft law document, Agenda 21, confirms this interpretation of the Declaration. Agenda 21 provides a comprehensive plan for strategies and programmes to reverse the effects of environmental degradation and promote sustainable development. Its text comprises four sections beginning with 'social and economic development' (including poverty alleviation, consumption patterns and integration) and 'conservation and management of resources for development' (including the atmosphere and biosphere).[128] As in the Declaration, the signatories of Agenda 21 agreed on a number of principles and concepts, but left a definition and the details of commitments of the developed world to further negotiations on treaties, institutions[129]

---

and environmental needs of present and future generations'.

125   Murswiek, D. (2002), 'Nachhaltigkeit – Probleme der rechtlichen Umsetzung eines umweltpolitischen Leitbildes', *Natur und Recht* 641–5.

126   See Sands 2003 (n. 68 above), 259. That the Rio Declaration did not replace or narrow down the principle of sustainability is also concluded, for example, by Murswiek 2002 (n. 125 above), 4–6; and Schmidt, R. and Müller, H. (2001), *Einführung in das Umweltrecht*, 6th edn (Munich, Beck), para. 7, 7.

127   Beyerlin, U. (1996), 'The Concept of Sustainable Development', in Wolfrum, R. (ed.), *Enforcing Environmental Standards: Economic Mechanisms as Viable Means?* (Berlin, Springer), 95, 104–5.

128   Section 3 deals with 'strengthening the role of major groups', section 4 with 'means of implementation'.

129   For example, the UN Commission for Sustainable Development (CSD) initiated and monitored national strategies for sustainable development and programmes for sustainability indicators to measure progress.

and laws. In the case of Agenda 21, this expectation was not unreasonable. At least, civil society, particularly local governments and educational institutions, used Agenda 21 as a genuine blueprint for actions towards sustainable development.

The general legacy of Rio, however, is one of unfinished business. With the absence of a binding treaty and clear definitions, an important opportunity was lost. In the closing statement of UNCED, Secretary General Maurice Strong said: 'We have a profoundly important *Declaration*, but it must continue to evolve towards what many of us hope will be an *Earth Charter* that could be finally sanctioned on the fiftieth anniversary of the United Nations in 1995'.[130] Maurice Strong went on to initiate a worldwide campaign for an Earth Charter. Created by several thousand civil society groups, but without direct input from states, the Earth Charter was eventually launched at the Peace Palace in The Hague in 2000. It represents a broader consensus on the principle of sustainability than has ever been achieved before. It was now up to the states to respond to the challenge.

Shortly after the launch, more than a thousand non-governmental organizations (NGOs) at the Millennium NGO Forum endorsed the Earth Charter and recommended that the UN Millennium Summit recognize and support the document. While this did not happen, the UN Millennium Declaration[131] does, for the first time in two decades, reaffirm the principle of 'respect for nature' as among the 'fundamental values essential to international relations'.[132] It also identifies freedom, equality, solidarity, tolerance and shared responsibility as fundamental values and calls for a 'new ethic'. In addition, the Millennium Development Goals[133] associated with the declaration are entirely consistent with the Earth Charter.[134]

Then came the 2002 World Summit on Sustainable Development (WSSD) in Johannesburg. At the opening, the Secretary General, South African President Mbeki, referred to the Earth Charter as a central document to guide the Summit's negotiations. The draft Johannesburg Declaration contained specific reference to the Charter and called for a commitment to the values and principles outlined there. However, the final version had this reference removed following last-minute objections, mainly from the United States.[135] Like ten years earlier, before UNCED, several PrepComs to the WSSD had recognized the Earth Charter's key importance, and like ten years earlier, states again moved away from a firm commitment to the principle of sustainability.

---

130   Quoted by Vilela 2005 (n. 119 above), 18.

131   United Nations Millennium Declaration (GA Res. 55/2, UN GAOR, 55th session, 8th plenary meeting, UN Doc A/Res/55/2 (2000)).

132   The Earth Charter preamble reads: 'We must join together to bring forth a sustainable society founded on respect for nature, universal human rights, economic justice, and a culture of peace'.

133   See Millennium Declaration (n. 131 above).

134   Rockefeller, S. (2005), 'The Transition to Sustainability', in Blaze Corcoran, P. (ed.), *The Earth Charter in Action. Toward a Sustainable World* (Amsterdam, KIT), 165, 167.

135   Ibid.

The key outcomes of the WSSD, the Johannesburg Declaration on Sustainable Development[136] and the Johannesburg Plan of Implementation,[137] failed to define sustainable development. They did, however, express support for the principle of ecological sustainability, albeit in an indirect manner. The Declaration commits to sustainable development and building a humane, equitable and caring global society cognizant of the need for human dignity for all. Paragraph 6 uses language almost identical to the Earth Charter's Preamble: 'From this continent, the cradle of humanity, we declare, through the Plan of Implementation of the World Summit on Sustainable Development and the present declaration, our responsibility to one another, to the greater community of life and to our children'. This statement is the first time that an international law document has made an explicit reference to the community of life. This affirmation deepens the meaning of respect for nature. At the same time, it is reflective not only of the Earth Charter's recognition that people are members of the Earth's community of life, but also of the recognition that non-human species are worthy of moral considerations. In other words, non-human species as members of the greater community of life have intrinsic value as well as instrumental value.[138]

While the Declaration is reflective of some of the ethics underpinning ecological sustainability, the Plan of Implementation acknowledges, for the first time, 'the importance of ethics for sustainable development'.[139] To this end, both soft law documents reveal a heightened sense of ecological responsibility. Their *travaux préparatoires* and actual texts do not signal an actual commitment to the principle of sustainability; however, they do signal a search for adopting it. If responsibilities are no longer limited to the social and economic side of sustainable development, but are inclusive of 'the greater community of life', then the formal recognition of the principle of sustainability becomes a distinct possibility. It all depends on how the 'ethics for sustainable development' are understood and applied.

Preceding the Johannesburg Summit was an effort by the International Law Association (ILA) to identify the key legal principles relevant to sustainable development. The 2002 New Delhi Declaration on the Principles of International Law Related to Sustainable Development[140] adopted seven such principles: (1) the duty of states to ensure sustainable use of natural resources; (2) the principle of equity and the eradication of poverty; (3) the principle of common but differentiated responsibilities; (4) the principle of the precautionary approach to human health, natural resources and ecosystems; (5) the principle of public participation and access to information and justice; (6) the principle of good governance; and (7) the principle of integration and interrelationship, in particular in relation to human rights and social, economic and environmental objectives.

---

136   Johannesburg Declaration on Sustainable Development (Johannesburg Declaration) (Johannesburg, 4 September 2002, UN Doc. A/Conf.199/20 (2002)).

137   Johannesburg Plan of Implementation (n. 70 above).

138   See Principles 1 and 15 of the Earth Charter; Rockefeller 2005 (n. 134 above), 167.

139   Johannesburg Plan of Implementation (n. 70 above), para. 6.

140   New Delhi Declaration on the Principles of International Law Related to Sustainable Development (London, 2002; ILA resolution 3/2002).

With respect to this last principle, the Declaration explains that it 'reflects the interdependence of social, economic, financial, environmental and human rights aspects of principles and rules of international law', and further that 'states should strive to resolve apparent conflict between competing economic, financial, social and environmental considerations' and finally that 'the above principles are interrelated and each of them should be construed in the context of the other principles of this Declaration'.[141] This suggests that there is no overall prime principle to guide the others. However, the first principle of sustainable use could be seen as having such a function if interpreted as a reflection of the principle of sustainability.

Content and limitations of the principle of sustainable use are uncertain under international law. On the one hand, there is the well-established right to permanent sovereignty over natural resources to which Principle 21 of the Stockholm Declaration and Principle 2 of the Rio Declaration added certain restrictions with respect to transboundary effects. On the other hand, the significance of this rule is far from clear. Commentators tend to give the duty of states not to harm the environment or, termed positively, the duty to manage natural resources in a sustainable manner, a certain prominence. Some consider this duty a 'central tenet of international environmental law',[142] while others speak of a 'nascent responsibility'.[143] As will be discussed later, the interpretation of the sustainable use depends on the ethics of sustainable development and the general importance of the principle of sustainability.

The period since the 2002 WSSD is marked by several international resolutions to endorse the Earth Charter. In 2003, the UNESCO General Conference recognized the Earth Charter as an important framework for sustainable development and affirmed member states' intention to use the Earth Charter 'as an educational instrument, particularly in the framework of the United Nations Decade on Education for Sustainable Development' (2005 to 2014).[144] The 2005 International Implementation Scheme for the UN Decade affirms that the Charter 'provides an excellent example of an inclusive vision of the fundamental principles for building a just, sustainable, and peaceful world'.[145]

In 2004, the IUCN World Conservation Congress adopted a resolution that 'endorses the Earth Charter as an inspirational expression of civil society's vision', 'recognises ... the Earth Charter as an ethical guide for IUCN policy', seeks to 'implement its principles through the IUCN Programme' and encourages member states 'to determine the role the Earth Charter can play as a policy guide within their

---

141   Ibid.

142   Segger and Khalfan 2004 (n. 69 above), 112 with reference to Sands 2003 (n. 68. above).

143   Shrijver, N. (1997), *Sovereignty over Natural Resources: Balancing Rights and Duties* (Cambridge, Cambridge University Press), 390–92.

144   UNESCO (2003), 'UNESCO's Support for the Earth Charter', in *Records of the General Conference 32nd session Paris, 29 September to 17 October 2003* (Paris, UNESCO), vol. 1, 35.

145   UNESCO, 'Education for Sustainable Development: Highlights on DESD Progress to Date', April 2007, 1, available at <www.portal.unesco.org/education/en/files/51172/11779357975Progress_to_Date_APRIL07.pdf/Progree+to+Date+APRIL07.pdf>, accessed 1 December 2007.

own spheres of responsibility'.[146] With only the United States expressly opposing the resolution and 67 of 77 member states plus 800 NGOs supporting it, the Earth Charter gained further recognition as expressing a consensus within the international community.

Such consensus does, at present, not amount to soft law recognition, nor is there a consensus among states on the Earth Charter's values and principles. However, the increased international recognition marks 'a significant, even if very gradual, shift in humanity's ethical awareness'.[147] Considering the ethical commitments of the Johannesburg documents and the recognized moral authority of the Earth Charter, the ethics of sustainability are more appreciated today than ten, fifteen years ago. It therefore becomes possible to adequately describe the core meaning of 'sustainable development'. A provisional description is to treat the idea of sustainability as the core reference for the concept of sustainable development. We will see later how the principle can also be defined as a fundamental principle of law.

Conclusion

Since 1972, and especially since 1992, the concept of sustainability seems to have lost its contours. Its popularization in the term 'sustainable development' created an invitation to use it for all sorts of objectives purported to be desirable ('sustainable economy', 'sustainable growth', 'sustainable policies', etc.). Many of such terminological constructs have no bearing to the original meaning of sustainability. If, for example, corporate managers stress economic sustainability as a legitimate goal next to ecological efficiency, chances are that they mean economic efficiency, but certainly not ecological sustainability. Economist Wilfred Beckerman was only slightly cynical when he wrote his article: 'How Would You Like Your "Sustainability", Sir? Weak or Strong?'[148]

But does the inflationary use of sustainability diminish its significance as a fundamental principle? Not if we focus on the ethical and legal context in which the term has been used. There is clear evidence to suggest that sustainability remains a distinct, relevant principle.

Following on from the 1972 Stockholm Declaration, the 1980 World Conservation Strategy and the 1982 World Charter of Nature, the 1987 Brundtland Report shaped its new concept of sustainable development around the old concept of sustainability. It did so by demanding global, long-term economic justice without sacrificing the Earth's ecological integrity. There may be many problems with a one-size-fit-all formula for solving the world's problems, but referring to sustainability is not one of them. It is the best term to capture ecological challenges that human societies, time and again, had to face.

---

146   International Union for the Conservation of Nature and Natural Resources *Endorsement of the Earth Charter* (Res. WCC 3.022 (2004)), available at <www.iucn.org/congress/resolutions>, accessed 1 December 2007.

147   Rockefeller 2005 (n. 134 above), 167.

148   Beckerman, W. (1995), ' "How Would You Like Your 'Sustainability', Sir? Weak or Strong?" A Reply to My Critics', *Environmental Values* 5, 169.

Ecological sustainability was implied in the Brundtland Report on sustainable development and, more importantly, in the 1992 Rio Declaration. While it is true that this document and the other more recent soft law documents on sustainable development contain political compromises and only general descriptions, it is also true that they are concerned with ecological sustainability. In fact, if today the concept of sustainable development is recognized as a principle of international law, it owes its main reason, namely operational quality, to the principle of sustainability. Without it, sustainable development (requiring the integration of environmental, social and economic goals) could not be made operable. This is critical for recognizing it as a legal norm. In other words, the concept of sustainable development can only perform its normative functions in so far as it incorporates the idea of ecological sustainability.

The continued existence of the principle of sustainability has two important consequences. The first is that sustainable development is given meaning and direction. For developed states there is no free choice between three equally relevant political objectives, but only one political goal: any use of natural resources has to be sustainable. Other goals like economic prosperity and social justice are secondary in a sense that they can only be pursued without threatening the Earth's ecological systems. Developing countries do not have the same responsibility. The principle of 'common but differentiated responsibilities', as defined in the Rio Declaration, affirmed in the Climate Change Convention and repeated six times in the Johannesburg Plan of Implementation, means that developed countries bear a special burden of responsibility for reducing and eliminating unsustainable patterns of production and consumption.[149]

The second consequence is that existing treaties, laws and legal principles need to be interpreted in the light of the principle of sustainability. It provides crucial guidance for the interpretation of legal norms and sets the benchmark for the understanding of justice, human rights and state sovereignty. In doing so, sustainability represents the foundational concept of emerging 'sustainability law' based on ecological justice, human rights and institutions.[150]

---

149 See Chapter 2.
150 See Chapters 3, 4 and 5.

Chapter 2

The Principle of Sustainability

What Constitutes a Legal Principle?

Fundamentally, law has a serving function. A legal system cannot on its own initiate and monitor social change; however, it can formulate some parameters for the direction and extent of social change. If such parameters are clear enough and reflective of what society feels strongly about, they will be effective. If they are unclear or ignorant to social realities, they will have little impact. It is crucial, therefore, to define those parameters clearly and realistically.

The parameters in the area of environmental law are those environmental principles that are legally recognized. This is not exclusively determined by lawyers, but by decision-makers in the broadest sense. Experts of environmental sciences belong to this group, but also those of philosophy and ethics, anthropology, economics, politics and other disciplines. They all contribute to what counts as a legally recognized environmental principle.

Environmental policy and law are charged with principles that originated in such an interdisciplinary context and eventually transformed into legal principles. Examples include the precautionary principle, the polluter pays principle, the principle of cooperation, the principle of integration, the principle of transparency and public participation, the principle of common but differentiated responsibility and the principle of sustainability.[1] They have in common that they are legally relevant and enforceable albeit not necessarily in the same manner and with equal importance.

But what act makes them 'legal' principles? There is no universal answer, as environmental law operates within different legal systems and at national, regional (EU) and international levels. Each legal system follows its own tradition to determine the legal nature of environmental principles. And while there is some commonality among environmental principles, no universal standard exists to define their exact legal nature.

An example is the precautionary principle, where even the term 'principle' is problematic. In some national jurisdictions the precautionary principle is a well-

---

1   Sadeleer, N. de (2002), *Environmental Principles* (Oxford, Oxford University Press), 23; Birnie, P. and Boyle, A. (2002), *International Law and the Environment*, 2nd edn (Oxford, Oxford University Press), 79; Sands, P. (2003), *Principles of International Environmental Law*, 2nd edn (Cambridge, Cambridge University Press), 231; Winter, G. (2004), 'The Legal Nature of Environmental Principles in International, EC and German Law', in Macroy, R. (ed.), *Principles of European Environmental Law* (Groningen, Europa Law Publishing) 9, 11.

established principle of environmental and public health law. Other jurisdictions know this 'principle' only as a broad 'approach' to guide environmental decision-making. These differences do not necessarily reflect varying degrees of defined content; in part, they reflect differences in legal cultures, for example, between the European civil law system and the Anglo-American common law system. At the level of the European Union law, the precautionary principle has received constitutional recognition,[2] but does not clearly distinguish between precaution and prevention as, for example, EU member state Germany does.[3] Its meaning is similar to the meaning that it has in international law. But even if the precautionary principle is considered to be part of customary international law,[4] a general principle[5] or even a rule,[6] the actual value of such classifications is limited. Where it is applied in cases of scientific uncertainty, its legal effect is open to broad interpretation. Does it require specific action to prevent harm or is any action (doing 'something') good enough? The example of the precautionary principle suggests precaution: there is no guarantee that a principle will work only because it has been recognized as a legal principle. Essentially, the importance of principles is not so much determined by their legal status, but by their interpretation through governments, courts and other decision-makers.

On the other hand, legal principles cause legal effects and can be enforced. Generally speaking, environmental principles are fairly similar around the world as they are responding to the same subject matter. They can, therefore, be identified and legally classified. As will be shown in the following, there are important differences between the various types of norms recognized in law. They differ in terms of their status, function and prescriptive nature. And there are also differences in terms of their operational context: principles of international environmental law operate at the level of the international community, while principles of domestic environmental law operate at the level of the national community. As a global concept, the principle of sustainability applies at both levels, but with different functions. The proper classification of the principle of sustainability is crucial as it will determine not only its legal nature, but also the general importance it may have within and between nations.

Definitions of 'principles' abound in the philosophical and legal literature. However, some basic classifications can be made for our purpose. A principle is commonly understood to have legal effect, if it is contained in a law. It does not matter how the law is created, whether by legislators, by courts or by other sources, for example, 'common law'. In international law, such law-creating factors are more complex than in domestic law. As there are no law-creating institutions comparable

---

2   Maastricht Treaty 1992, Article 174.

3   Krämer, L. (2004), 'The Genesis of EC Environmental Principles', in Macroy, R. (ed.), *Principles of European Environmental Law* (Groningen, Europa Law Publishing), 29, 39.

4   Stone, C. (2001), 'Is there a Precautionary Principle', *Environmental Law Reporter* 31, 10790.

5   Ibid.

6   Ibid.

to national parliaments or courts, international law is being created by a variety of sources or factors, many of which are not clearly defined. It is widely accepted that the traditional sources of international law (treaties, customary law, general principles) are insufficient to describe the spectrum of sources that are, in fact, law-creating. 'Soft law', for example, is an important source of international environmental law even though it lacks a legally binding quality.

The complexity of law-creating sources has caused some international lawyers to apply the term principle in a loose manner. According to them, an environmental principle does not need legal effect to make it relevant in international law. Astrid Epiney and M. Scheyli[7] consider the principles of sustainability, precaution and others as examples of this kind. They justify their validity for environmental law with their actual importance. International environmental law relies, in fact, on a number of general practices, concepts and ideas (regardless of their legal nature) that states take as guidance for their actions. Practice and trust are certainly important for the functioning and further development of international environmental law. Nevertheless, the line between law and non-law should not be blurred.[8] There are political, psychological and conceptual advantages in maintaining the distinction between, for example, moral principles and legal principles.

Accordingly, a principle that is not or not yet contained in law could be described as a moral principle, ideal, objective or policy[9] (depending on its context). Such a principle could well be referred to in a legally binding treaty, for example, the mentioning of 'sustainable development' in the Climate Change Convention,[10] but this does not in itself make sustainable development a legal principle. For this to happen, a law-creating act or process is necessary.

Sometimes a legal principle is created in a single act. This can be the case when it appears in constitutions and legislation or, in international law, in a treaty or landmark decision of the International Court of Justice.[11] However, more often the creation of legal principles is a process with many small steps. For example, when the precautionary principle appeared in the German Clean Air Act 1974, it did not automatically guide all environmental legislation. This happened through a process when other laws were gradually oriented towards precaution. Only in the summation of such laws did it become possible to identify precaution as a principle guiding German environmental law.

In international environmental law, the gradual emergence of recognized principles is the absolute rule. Virtually all its general principles and rules emerged

---

7   Epiney, A. and Scheyli, M. (2000), *Umweltvölkerrecht* (Bern, Stämpfli), 75; See also Epiney, A. and Scheyli, M. (1998), *Strukturprinzipien des Umweltvölkerrechts* (Bern, Stämpfli), 171 and, with respect to the precautionary principle, Hohmann, H. (1994), *Precautionary Legal Duties and Principles of Modern International and Environmental Law* (London and Boston, Graham & Trotman), 166.

8   See further below.

9   For a distinction between ideals and policies see Verschuuren, J. (2003), *Principles of Environmental Law* (Baden-Baden, Nomos), 19.

10   United Nations Framework Convention on Climate Change (New York, 9 May 1992; UN Doc. A/AC.237/18, 31 ILM 848), Article 3 (4).

11   For example, the good-neighbourliness principle.

incrementally, over a certain period of time and not without controversy.[12] Whether or not a principle is recognized through treaty law or meets the requirements for a 'general principle' or 'customary rule'[13] is often not clear, but rather a matter of criteria being used. And much depends on the persuasiveness of the principle itself. If it reflects common sense or common practice, then its chances for legal recognition are high. If it is merely being promoted through repetition, then legal recognition is less likely, especially if its exact meaning remains unclear.

In conclusion, principles such as the principle of sustainability or any other environmental principle gain validity as soon as they are recognized as sufficiently relevant. They can influence policies and laws quite independently of their legal nature. However, in order to be recognized within law, they have to be given legal effect. This can happen in many ways, for example, through a single act or incrementally through legal institutions, states' behaviour or demonstrated international consensus. The principle of sustainability assumes its validity through long-standing practice and public awareness. Whether or not it is a legal principle depends on the classification system used in national and international environmental law.

A Typology of Environmental Legal Principles

There are different types of legal norms. Their respective characteristics need to be clarified to distinguish their legal meaning, but also to distinguish them from non-legal norms. A typology of (non-legal and legal) environmental principles can help to determine the legal nature of the principle of sustainability.

As mentioned earlier, environmental law is built around environmental principles that originated partly in law and partly in other disciplines including ethics, science, economics as well as foundational cultural concepts. Essentially, we can perceive environmental law as an interdisciplinary field with many disciplines contributing to its characteristics.[14] For this reason, it is often difficult to determine the legal nature of relevant principles. Sometimes, a scientific concept like, for example, 'biological diversity' creates an entirely new area of environmental legislation, whereas a legal principle such as 'polluter pays' may have very little or even no impact on environmental law.[15] The great variety of relevant principles is, however, no reason to ignore the distinction between legal and non-legal norms. On the contrary, it stresses the need for proper definitions. No matter how influential non-legal concepts may

---

12   On the one hand, the principles such as precaution, cooperation, polluter pays or sustainable use are widely accepted as legal principles, yet their exact classification – as general principle, rule or custom – continues to be disputed.

13   The traditional sources of international law are listed in the Statute of the International Court of Justice, Article 38.

14   Bosselmann, K. and Grinlinton. D. (eds) (2002), *Environmental Law for a Sustainable Society* (Auckland, New Zealand Centre for Environmental Law), vol. 1, vii–ix; Kiss, A. and Shelton, D. (2004), *International Environmental Law*, 3rd edn (New York, Transnational Publishers), 1–26.

15   Sands 2003 (n. 1 above), 279.

be in practice, only legal norms have *direct* effect for legal processes, instruments and outcomes.

If environmental law is in need of accurate norm classification, this is particularly so for international environmental law. Sharing many of the normative uncertainties of international law, the field of international environmental law is burdened with the additional uncertainties of environmental law as just mentioned. As a result, we are faced with hardly ever knowing the exact status of a particular principle. The rule seems uncertainty and clarity the exception.

To make matters worse, in between non-legal and legal norms international lawyers have identified 'relative normativity'[16] which Prosper Weil described as a 'pathological phenomenon'[17] as it sits squarely with the idea of legal certainty.

In an attempt to clarify the status of what he calls 'twilight' norms, Ulrich Beyerlin[18] develops a typology that takes account of both, the need for separating legal norms from non-legal norms and the need for giving due recognition to emerging principles such as, for example, sustainable development.

The Difference between Policies and Principles

Beyerlin defines as 'twilight' any norm that does not clearly enough set out legal consequences that follow automatically when the facts it stipulates are given. This definition goes back to Ronald Dworkin's understanding of what legal rules should be. Dworkin separates 'policies' from 'legal principles' and 'legal rules'.[19] He draws what he calls a 'logical' distinction between legal principles and legal rules: 'Both sets of standards point to particular decisions about legal obligation in particular circumstances, but they differ in the character of direction they give'. A legal rule applies in 'an all or nothing fashion', although exceptions may be possible, while a legal principle 'states a reason that argues in one direction, but does not necessitate a particular decision'.[20]

A further consideration of Dworkin's distinction between principles and rules is that, given the need to choose between different principles for reaching a decision, it may be necessary to consider their weight or importance. This would suggest a hierarchy between principles and also that a principle can be more important than a rule. According to Dworkin, the difference between a principle and a rule is that a rule (normally) leads to a particular consequence. Thus far, Beyerlin follows Dworkin by asserting that twilight norms are even less clear in terms of leading to a particular

---

16   Beyerlin, U. (2007), 'Different Types of Norms in International Environmental Law: Policies, Principles and Rules', in Bodansky, D., Brunnée, J. and Hey, E. (eds), *Oxford Handbook of International Environmental Law* (Oxford, Oxford University Press), Chapter 18, with reference to Weil, P. (1983), 'Towards Relative Normativity in International Law?', *AJIL* 77, 413–42 and Fastenrath, U. (1993), 'Relative Normativity in International Law', *EJIL* 4:1, 305–40.

17   Weil 1983 (n. 16 above), 414.

18   Beyerlin 2007 (n. 16 above).

19   Dworkin, R. (1977), *Taking Rights Seriously* (Cambridge, MA, Harvard University Press), 22.

20   Ibid. Chapter 2.

consequence, but that this says nothing about their actual importance. The concept of sustainable development, therefore, would not be diminished in importance if not classified as a legal norm.[21]

Dworkin identified his three types of norms in response to H.L.A Hart's legal positivism that only allowed for a 'rule', thereby neglecting 'the important roles of these standards that are not rules'.[22] Dworkin's greater flexibility fits international environmental law well, as it allows for closer linkages between policies and principles without sacrificing the fundamental difference between non-legal and legal principles.

This is visible in his definitions for 'policy', on the one hand, and 'principle' on the other. He understands 'policy' as 'that kind of standard that sets out a goal to be reached, generally an improvement in some economic, political, or social feature of the community'. A 'principle' is seen as 'a standard that is to be observed, not because it will advance or secure an economic, or social situation deemed desirable, but because it is a requirement of justice or fairness or some other dimension of morality'.[23] These definitions make it clear why policies do not possess the same legal potential as principles. The differentiating factor is that a principle is derived from a more fundamental concern such as justice, fairness or some other moral principle that law is ultimately grounded in. This explains convincingly the importance of separating non-legal from legal principles. So long as policies (ideals, objectives, etc.), no matter how desirable, are not grounded in some legally recognized morality, they are not candidates for becoming legal principles. However, if they are so grounded, they can attain the level of a legal principle.

There is a lot of support among international environmental lawyers for the idea that morality is of foundational importance. According to Alexander Kiss and Dinah Shelton, international environmental law is largely shaped through the 'religious, ethical and philosophical foundations of environmental protection'.[24] 'Religious and philosophical concepts are crucial to understanding the views of nature and humankind's relationship to it that form the basis of environmental law'.[25] In the second edition of his seminal text on *Principles of International Environmental Law* Philippe Sands says: 'Since the first edition of the book, greater attention has ... been given to other values, representing neither scientific nor economic considerations'.[26] He explains these 'other values' in the context of EC rules on genetic engineering, the 2000 Biosafety Protocol with its reference to 'the value of biological diversity' and the WTO ruling in the 2000 *Asbestos* case where it was recognized that the characteristics of a consumer product go beyond economical, physical and scientific considerations.[27] Sands goes on to present the concept of sustainable development as a substantial broadening of environmental law, now including moral issues such as

---

21   This is the conclusion Beyerlin reaches; Beyerlin 2007 (n. 16 above), 20.
22   Dworkin 1977 (n. 19 above), 22.
23   Ibid.
24   Kiss and Shelton 2004 (n. 14 above), 11.
25   Ibid.
26   Sands 2003 (n. 1 above), 6.
27   Ibid. 10.

intra- and intergenerational equity.[28] The ethical nature of the concept of sustainable development is widely accepted,[29] but its legal ramifications in the formation of 'sustainable development law'[30] are yet to be recognized.

If environmental ethics are applied to environmental law, the issue of morality becomes a legal challenge. We can ask, for example, how environmental law could ever make a contribution to protecting the environment, if its own conceptual foundations are too weak. However, the challenge is not to realize current shortcomings such as, for example, anthropocentric reductionism, but the actual integration of the ethical discourse with the legal discourse. As an environmental principle hardly ever emerges from an ethical vacuum, its morality cannot be ignored. If its moral dimension is resonant with a legal dimension, then there is reason to recognize its legal nature.

This is behind Dworkin's definition of a principle. His reference to justice, fairness and other forms of foundational morality is a reminder of the law's own roots. Only a legal positivist can ignore that law needs to find its legitimacy in foundational concepts like justice, fairness or equality. For abstract policies, ideals or objectives to gain recognition as legal principles they need to represent requirements of foundational legal morality.

According to Ronald Dworkin, 'policies' are separated from 'principles' also by their different addressees: 'Arguments of principle are arguments intended to establish an individual right; arguments of policy are arguments intended to establish a collective goal. Principles are propositions that describe rights; policies are propositions that describe goals'.[31] Such separation does not, however, reflect legal reality. There are many 'principles' that are geared towards public interests rather than individual rights. Environmental principles are a prime example. In instances like the environment or public health, individual interests cannot be separated from public interests. The actual content is, therefore, not what separates principles from policies, it is purely their legal potential, that is, their closeness to existing legal principles or fundamental legal norms.[32]

Dworkin's characterization of a principle as having both moral and legal significance seems to be widely accepted among international lawyers, although there is little discussion about this particular aspect. A. Aarnio is critical of Dworkin's dichotomy between principles and rules,[33] suggesting four different categories

---

28 Ibid. 10–12.

29 Not least in the Johannesburg Plan of Implementation (Johannesburg, 4 September 2002; UN Doc. A/Conf.199/20).

30 Cordonier Segger, M.C. and Khalfan, A. (2004), *Sustainable Development Law: Principles, Practices and Prospects* (Oxford, Oxford University Press). For a book claiming to 'provide a current guide to this emerging area, a framework and methodology for researching and implementing sustainable development law' (365) the absence of the conceptual debate surrounding sustainable development, such as the 'weak' vs 'strong' controversy, is a major omission, although the authors admit that 'this may be the path less travelled' (372).

31 Dworkin 1977 (n. 19 above), 90.

32 Winter 2004 (n. 1 above), 14.

33 Beyerlin 2007 (n. 16 above); Winter 2004 (n. 1 above).

instead.[34] He describes them as 'rules proper', 'rule-like principles', 'principle-like rules' and 'principles proper'.[35] However, his explanation of 'principles proper' is similar to Dworkin's definition noting 'equality, liberty and other value principles and goal principles' as examples.

With respect to international environmental law, Nicholas de Sadeleer notes that the creation of relevant principles is less achieved by courts or treaty law, but by bold statements in both hard law and soft law. The reason for this greater flexibility of law-creation is the wider acceptance of principles. Core environmental principles, he suggests, provide a solid coherence and stability in a new legal area that needs to be adaptable and fast-moving.[36] He refers to them as 'the interface between modern law and post modern law'[37] as they have important 'enabling',[38] 'directing'[39] and 'interpretative'[40] functions for legal decision-making. De Sadeleer asserts that these various functions of environmental principles 'weaken the dichotomy put forward by Dworkin',[41] but agrees with him fundamentally. Moral principles possess law-creating potential, especially in environmental law.

Sustainable Development as a Legal Principle

Vaughan Lowe holds similar views in his characterization of sustainable development.[42] He shares the widely held view that there are normative aspects associated with this concept and that they ought to be recognized by law. His approach, locating sustainable development somewhere between (non-legal) policies and (legal) principles is helpful, not because of its thesis, but because it helps in clarifying the issues.

Lowe considers sustainable development as 'a meta-principle, acting upon other legal rules and principles – a legal concept exercising a kind of interstitial normativity, pushing and pulling the boundaries of true primary norms when they threaten to overlap or conflict with each other'.[43] This description differs

---

34   Aarnio, A. (1990), 'Taking Rules Seriously', *ARSP* 42, 180–92.

35   Ibid. 184.

36   Sadeleer, N. de (2004), 'Environmental Principles, Modern and Post-Modern Law', in Macroy, R. (ed.), *Principles of European Environmental Law* (Groningen, Europa Law Publishing), 225–36, 231; see also Macroy, R. (2004), 'Principles into Practice', in Macroy, R. (ed.), *Principles of European Environmental Law* (Groningen, Europa Law Publishing) 1–8, 4.

37   Sadeleer 2004 (n. 36 above), 231.

38   Ibid. 232

39   Ibid. 233.

40   Ibid. 234.

41   Sadeleer 2002 (n. 1above), 308.

42   Lowe, V. (1999), 'Sustainable Development and Unsustainable Arguments', in Boyle, A. and Freestone, D. (eds), *International Law and Sustainable Development: Past Achievements and Future Challenges* (Oxford, Oxford University Press), 19–37, 19.

43   Ibid. 31.

from Sands's interpretation of the *Gabçikovo-Nagymaros* case[44] that Lowe finds 'not sustainable'.[45] According to Sands there 'can be little doubt that the concept of "sustainable development" has entered the corpus of international customary law'.[46] Lowe, on the other hand, insists that sustainable development 'is itself not a norm; it can be no more than a name for a set of norms. Indeed, it may not even be that'.[47] Underlying this critique is Lowe's concern for the normative quality of principles. According to him, sustainable development may be 'pushing and pulling the boundaries of true primary norms', but cannot itself provide the legal guidance typical for norms and principles.

These observations point to a dilemma by which the entire sustainable development debate is hampered. On the one hand, there is a lot of sympathy among international environmental lawyers and activists for giving sustainable development 'legal teeth'. After all, states have not been forthcoming in implementing it. On the other hand, many feel that sustainable development is too vague a concept, an empty shell precisely because it lacks the political and legal guidance necessary for implementation. It cannot be both politically ambiguous *and* legally effective.

Lowe's idea of a 'meta-principle' that has 'interstitial normativity' and is capable of 'modifying norms' may be appealing to some as it seemingly bridges the gap between political vagueness and legal effect. However, it does not solve the aforementioned dilemma and is not helpful in practical terms. If states take political guidance from sustainable development, they can do so regardless of its legal nature. If, on the other hand, states would take that guidance only if required by law, the hybrid 'meta-principle' would not achieve that. Only principles can amount to legal obligations; a 'meta-principle' can, at best, help explaining them. There is little point in blurring the boundaries between policies and principles if nothing can be gained from it.

There have been other attempts, like Lowe's, to blur the boundaries. As mentioned,[48] Epiney and Scheyli identified new types of general principles to describe the practical functioning of international environmental law. They may create a kind of proto-law guiding or influencing 'real' law, but do not explain why some principles are, in fact, treated as legal principles along with other legal norms such as rules. Moreover, the emergence of new principles out of common sense and good practice could be hampered. If the discourse about new principles is burdened with the 'threat' that common sense alone makes applicable law, we may not see their emergence at all. Gerd Winter rightly makes the point that the dynamic potential of principles is based on their somewhat elusive status behind the scenes.[49]

For these reasons, it is not unproblematic when Wilfred Lang proposes 'three different categories of principles of a decreasing legally binding/compulsory

---

44   *Case Concerning the Gabçikovo-Nagmaros Project (Hungry/Slovakia)*, 1997 ICJ, 37 ILM 162 (1998).
45   Lowe 1999 (n. 42 above), 30.
46   Sands 2003 (n. 1 above), 254.
47   Lowe 1999 (n. 42 above), 31.
48   Epiney and Scheyli 2000 (n. 7 above) and Epiney and Scheyli 1998 (n. 7 above).
49   Winter 2004 (n. 1 above), 18 at Footnote 25.

nature'.[50] He separates 'principles of existing International Environmental Law' from 'principles of emerging International Environmental Law' and 'potential principles of International Environmental Law'. Lang sees this last category as 'an area of hope for many policy-makers'[51] and asks where the boundary between policy and law is. His observation in this regard – 'customary law is a highly difficult and complex area of law'[52] – is, however, no answer.

Where does this leave us with sustainable development? If this concept is caught between political vagueness and legal ambition, there are only two ways out of the dilemma. International lawyers could either conclude that the composite term 'sustainable development' is an empty phrase, thus useless and not binding anyone;[53] or they could be serious about detecting its normative core. This latter option should be followed; it is the only one available to make sustainable development operable and enforceable.

To identify the normative core is not difficult, at least from a logical point of view. If 'sustainable' qualifies 'development' as distinguished from unsustainable development, its meaning must be clear. If, on the other hand, the attribute 'sustainable' remains undefined, the direction away from 'unsustainable' cannot be found, which means that no political guidance could be provided and no legal imperatives could follow. It should also be noted that the necessary clarity cannot be gained by merely referring to social, economic and environmental goals as all being relevant to sustainable development. Nor does it help to consider these goals as equally important. To identify, and act upon, potential conflicts between them, it is necessary to know how they should be resolved. If, for example, economic interests are in conflict with environmental interests, as often will be the case, we cannot rely on the expectation that existing legal principles such as the precautionary principle will lead the decision-making process in the right direction. The reason is that they are in themselves not ethically grounded in the same way as the principle of sustainability.[54]

---

50   Lang, W. (1999), 'UN Principles and International Environmental Law', *Max Planck Yearbook of United Nations Law* 3, 157–72, 171.

51   Ibid.

52   Ibid. 170.

53   It does not help to describe 'sustainable development' as a mere 'ideal' or 'value' as Jonathan Verschuuren suggests; Verschuuren 2003 (n. 9 above), 20. While acknowledging that 'an ideal is a value that is explicit, implicit, or latent in the law' (49), his interpretation of sustainable development as inevitably anthropocentric leads him to the conclusion that it has no normative quality and cannot be a legal principle. Why should the perceived anthropocentric character be a reason to deny 'sustainable development' legal quality? The (preferred or rejected) ethical foundation of an ideal – whether anthropocentric or ecocentric – can hardly be a criterion for its legal validity. What matters is its norm-creating potential: does it provide clear enough guidance or not?

54   This is overlooked by Jonathan Verschuuren; Verschuuren 2003 (n. 9 above), 50. He assumes an eco-ethical core ('room') in existing principles and processes that is simply not there ('The ecological aspects of the ideal of sustainable development can be sufficiently advanced in decision-making processes by governmental authorities and courts, because most

Clarity can only come from defining the essence of 'sustainable' with respect to its object. The essence is neither 'economic sustainability', nor 'social sustainability', nor 'everything sustainable', but 'ecological sustainability'. This is not the same as saying economic and social goals are less important. Both are an integral part of the concept of sustainable development, but they are not an integral part of the principle of sustainability. Rolling the three forms of sustainability into one principle would be impossible without giving up its core meaning. In other words, it can only be either 'economic sustainability' (whatever its meaning) *or* 'social sustainability' *or* 'ecological sustainability' that can provide the necessary direction.

So much for the logical argument. The conceptual argument is that the principle of sustainability has been in existence for centuries with never any other object than the natural resource base. While this object may have been broadened – from local resources in ancient times to ecosystems in recent times to planetary ecosystems today – the principle of ecological sustainability never changed. This core of sustainability cannot be different from 'sustainable' in the context of 'development'. The fact that social and economic aspects are included in the concept of 'sustainable development' does, therefore, not require any deviation from the ecological core. On the contrary, only because of this is it possible to relate the social and economic components of sustainable development to a central point of reference. As a consequence, the entire concept becomes operable: development is sustainable if it tends to preserve the integrity and continued existence of ecological systems; it is unsustainable if it tends to do otherwise.

This holistic, yet structured, concept of sustainable development equals 'ecologically sustainable development' and can be interpreted in the following way: 'No economic prosperity without social justice and no social justice without economic prosperity, and both within the limits of ecological sustainability'. As a norm this can be formulated as the obligation to promote long-term economic prosperity and social justice within the limits of ecological sustainability.

The principle of sustainability itself is best defined as the duty to protect and restore the integrity of the Earth's ecological systems. The further contents and dimensions of the principle will be discussed further below. For our present purpose it should become clear that the principle has a normative quality. It is reflective of a fundamental morality (respect for ecological integrity), requires action ('protect and restore') and is, therefore, capable of causing legal effect. The standards of a legal principle are all met.

With respect to the concept of sustainable development, the principle provides important guidance to make the concept operable. Whether this amounts to determinable legal content, making sustainable development a legal principle, is a matter of debate. In what follows, I will argue in favour of a legal principle, but one that has not yet been recognized as such by international law.

Before doing so, we need to briefly discuss the difference between a principle and a rule, not because sustainable development might be considered a 'rule' – it is clearly not one – but because it helps to further determine the characteristics of

principles that rule environmental decision-making processes create enough room to take into account the more eco-centric arguments').

a 'principle'. Principles and rules are equally valid legal norms, but with different functions. As mentioned earlier,[55] Dworkin has defined rules as setting direct legal consequences. If the facts that a rule stipulates are present, then specific measures are to be taken. Principles, by contrast, 'do not set out legal consequences that follow automatically when the conditions provided are met'.[56] While this separation is widely accepted as a rough guide, there are variations. Sometimes, rules can allow for exceptions. On other occasions, principles can be 'rule-like'[57] in a sense that they are uncompromising and absolute. Human dignity, for example, cannot be compromised by other principles, no matter how important they are. International environmental law is built around such flexibility of rules and principles, which does not mean, however, that any distinctions between them are obsolete.

Most legal theorists hold the view that principles differ from rules by degrees, rather than substantially,[58] with principles representing 'a greater generality than rules'[59] and rules having more focus and practicality. Sands describes rules as the 'practical formulation of the principles'[60] referring to a classic case[61] and Dworkin's concern for applying rules 'in an all-or-nothing fashion'.[62] Furthermore, principles can appear as 'rules of indeterminate content'[63] with a degree of abstraction that makes it impossible to deduce specific obligations, including rules, from them. They can, nevertheless, be of such high importance that they may constitute foundational principles (e.g. equality or legal certainty) to guide all other principles and norms.[64]

The concept of sustainable development is located somewhere in the spectrum of (non-legal) policies and (legal) norms. We can perhaps exclude two extremes. On the one hand, the concept is more than a mere political ideal as it has been referred to repeatedly in hard law[65] and, in much detail, in soft law. On the other hand, the concept is less than a rule as it lacks direct legal consequence. The proximity to policies seems closer than to a rule, but how close? This depends on the degree of political and legal guidance the concept is capable of providing.

---

55   Dworkin 1977 (n. 19 above).

56   Ibid. 24.

57   Aarnio 1990 (n. 34 above), 181.

58   MacCormick, N. (1978), *Legal Reasoning and Legal Theory* (repr. 1994, Oxford, Clarendon Press), 152; Raz, L. (1975), *Practical Reason and Norms* (repr. 1990, London, Hutchinson), 48.

59   Aarnio 1990 (n. 34 above), 181. See also Verschuuren 2003 (n. 9 above), 40.

60   Sands 2003 (n. 1 above), 233.

61   Gentini case (*Italy/Venezuela*) MCC (1903), 'a rule is essentially practical', while a principle 'expresses a general truth'; cited in Cheng, B. (1953), *General Principles of Law as Applied by International Courts and Tribunals* (London, Stevens), 376.

62   Dworkin 1977 (n. 19 above), 24.

63   Kiss and Shelton 2004 (n. 14 above), 203.

64   Ibid.

65   Climate Change Convention 1992 (n. 10 above), Article 3; Biodiversity Convention 1992 (Rio de Janiero, 5 June 1992; 31 ILM 818 (1992)), Articles 8, 10; Desertification Convention 1997, Articles 4, 5; Antigua Convention 1992, Article 3; For further examples see Cordonier Segger and Khalfan 2004 (n. 30 above), 95.

The composite term 'sustainable development' itself provides little guidance as it could be interpreted either way, that is, as essentially referring to ecological sustainability or as describing an undefined mix of environmental, social and economic goals to achieve sustainable development. This ambiguity would not allow the conclusion that the term has any tangible steering effect on states' environmental behaviour.

If not the term, perhaps the concept has such a steering effect? The concept is visible in some global and regional treaties and the explanations in the Brundtland Report, the Rio Declaration, Agenda 21, the Millennium Development Goals, the Johannesburg Declaration and the Johannesburg Plan of Implementation. On this basis, the concept can be described in a concise manner. Famous examples include the Brundtland definition with its emphasis on 'needs' or Article 4 of the Rio Declaration with its emphasis on 'integration': 'In order to achieve sustainable development, environmental protection shall constitute an integral part of the development process and cannot be considered in isolation from it'. Profiled in such a way, sustainable development may be understood as a normative concept with sufficient steering effect for states. The evidence for this can probably not be found by merely citing the concept or its definitions, but certainly by analysing its history and ethical dimension.

A proper analysis of the concept of sustainable development needs to include the entire history of its emergence, the various cultural contexts in which it appeared and the environmental jurisprudence of international courts, as will be discussed further, below. Such deeper analysis can detect the evidence for ecological sustainability as the core idea of the concept. There can be little doubt that the principle of sustainability has shaped the concept, at least historically, then with its emergence in the 1980s and right through to the present time.

As shown in the first chapter, the ecological bottom line that the principle represents was noted in the 1992 Rio Declaration. It has informed the Declaration's description of sustainable development by giving direction, in particular, to the principle's integration and common, but differentiated, responsibility. The protection and preservation of the natural resource base sets the benchmark against which the social and economic dimensions of sustainable development are to be achieved. Rich ('developed') countries bear a special responsibility to eliminate unsustainable patterns of production and consumption to facilitate sustainable development in poor countries. The normative value of the concept of sustainable development may not be as clear as in many other principles, but this is owing to the complexity of issues involved, not to a normative ambiguity of the concept itself.

Further evidence for the normative character of sustainable development can be found in the Johannesburg Declaration and Plan of Implementation. With respect to sustainable development, the Declaration acknowledges 'our responsibility to one another, to the greater community of life and our children',[66] while the

---

66 Johannesburg Declaration on Sustainable Development (Johannesburg Declaration) (Johannesburg, 4 September 2002, UN Doc. A/Conf.199/20 (2002)), paragraph 6.

Plan of Implementation acknowledges 'the importance of ethics for sustainable development'.[67]

The ICJ *Case Concerning the Gabçikovo-Nagymaros Dam (Hungary/Slovakia)* of 1997 is also important here. Both parties invoked the concept of sustainable development. Slovakia stated that: 'It is clear from both the letter and the spirit of these principles ... that development should proceed *in a way that is environmentally sustainable*'.[68] Hungary replied: 'Well-established ... operational concepts like 'sustainable development' ... help define, in particular cases, the basis upon which to assess the legality of actions such as the unilateral diversion of the Danube by Czechoslovakia and its continuation by Slovakia'.[69] The ICJ agreed that the 'need to reconcile economic development with protection of the environment is aptly expressed in the concept of sustainable development'[70] and added: 'For the purposes of the present case, this means that the Parties together should look afresh at the effects on the environment of the operation of the Gabcikovo power plant'.[71] While the Court stopped short of saying that sustainable development 'is a principle with normative value',[72] it did certainly confirm its guiding function, which Sands interprets as the Court's recognition 'that the term has a legal function'.[73]

The above analysis shows that the principle of sustainability cannot be overlooked in the various contexts where the concept of sustainable development has been used. The key documents and judicial authorities of international law have implicitly recognized its existence in their interpretation of sustainable development. In the words of former ICJ Vice President Weeramantry: 'It is thus the correct formulation of the right to development that that right does not exist in the absolute sense, but is relative always to its tolerance by the environment'.[74]

Sustainable development has, therefore, sufficient normative character to justify its classification as a legal principle. Whether this also means that sustainable development can today be seen as forming part of international customary law is a different matter. There is, of course, no automatic link between the classification of the concept as a legal principle and its full recognition in international law. On the other hand, commentators often question its international legal status not so much on the basis of lacking general practice by states or *opinio iuris*, but more in view of its perceived ambiguity and vagueness.[75] With the clarification that the concept

---

67   Johannesburg Plan of Implementation (Johannesburg, 4 September 2002, UN Doc. A/Conf.199/20), Article 6.

68   Quoted from Sands 2003 (n. 1 above), 255 at Footnote 118 (emphasis added).

69   Ibid.

70   *Gabcikovo-Nagymaros Case* 1998 (n. 44 above).

71   Ibid.

72   Ibid. This is the classification by Judge Weeramantry in his Separate Opinion.

73   Sands 2003 (n. 1 above), 254.

74   *Gabcikovo-Nagymaros Case* 1998 (n. 44 above), Judge Weeramantry's Separate Opinion.

75   See, for example, Beyerlin, U. (1996), 'The Concept of Sustainable Development', in Wolfrum, R. (ed.), *Enforcing Environmental Standards: Economic Mechanisms as Viable Means?* (Berlin, Springer), 95–121; Beyerlin 2007 (n. 16 above), 20; Winter 2004 (n. 1 above), 14; Lowe 1999 (n. 42 above), 30.

gains from utilizing the principle of sustainability, such concerns can be met. There is certainly no reason to categorically deny the ability of sustainable development to gain international legal status.

It has to be said, though, that the principle of sustainable development has not yet gained full international legal status, as has, for example, the precautionary principle. It has not been included in an international treaty, nor has it been accepted in the general practice of states accompanied by *opinio iuris*. However, this is hardly surprising. The political and legal potential of the underlying principle of sustainability has yet to be fully explored by the jurisprudence of international law.

Sustainability as a Fundamental Principle of Law

Our deliberations on the classification of policies and principles have shown that sustainable development is a legal principle.[76] This was explained with the normative character that the principle of sustainability has for the meaning of sustainable development. It follows that sustainability possesses the quality of a legal principle. We have defined it as the duty to protect and restore the integrity of the Earth's ecological systems.

In this section we will explore this definition further. We will first relate it to similar duties promoted in international environmental law. This should also clarify the relationship of the principle of sustainability to other environmental principles, whether they are recognized as legal norms or not. We will then locate sustainability as the most fundamental environmental principle, only equal to other fundamental principles of law such as freedom, equality and justice. The most profound and important document to recognize the fundamental importance of sustainability is the Earth Charter. We will see how much the Earth Charter relies on this principle in order to define a framework for building a just, sustainable and peaceful world.

The Relationship to Other Environmental Principles

Compared to other areas of international law, international environmental law is severely underdeveloped. It has no globally binding instrument that sets out the rights and duties of states with respect to the environment. By contrast, international human rights law,[77] international labour law[78] or international trade law[79] are each guided by global treaties containing fundamental rights and obligations. Considering that these areas contribute substantially to the success (or failure) of sustainable development, the absence of a foundational treaty for the protection of the global

---

76   In agreement with Tladi, D. (2007), *Sustainable Development in International Law* (Pretoria, Pretoria University Press), 112.

77   For example, Universal Declaration of Human Rights 1949; International Covenant on Civil and Political Rights 1966; International Covenant on Economic, Social and Cultural Rights 1966.

78   For example, International Labor Convention 1949.

79   For example, Marrakesh Agreement Establishing the World Trade Organization 1994; General Agreement on Trade in Services 1994.

environment is significant. Essentially, this means that environmental rights and obligations are not codified. The only exceptions are obligations with respect to specific environmental problems (such as climate change or biodiversity). On the other hand, the right to use and exploit the environment is an integral part of the concept of state sovereignty, the most important principle of international law. The rudimentary state of international environmental law is one of the reasons why there is, at present, no 'international sustainable development law' as a coherent body of law.

Instead of treaty obligations to protect the environment, we only have 'principles' and 'rules'. They are not fixed and clearly defined, as we have seen, but need to be derived from assessments of state practice, soft law documents, decisions of courts and tribunals, common sense and the like. In the absence of a central legal authority, the global environment has to rely on these sources and hope that states and global civil society take their obligations seriously.

The rules and principles of international environmental law are not laid down as such, but they have been formulated and promoted by numerous international commissions, working groups and initiatives as well as soft law declarations of states. Examples include the Stockholm Declaration (1972), World Commission on Environment and Development (1987), the Rio Declaration (1992), the IUCN Draft Covenant on Environment and Development (1995), the Earth Charter (2000) and the International Law Association's New Delhi Declaration of Principles of International Law relating to Sustainable Development (2002). The last three documents are of particular interest. They are particularly reflective of sustainability-related principles. While the IUCN Draft Covenant and the Earth Charter are fully developed 'constitutions', the New Delhi Declaration defines some (not all) principles in support of sustainable development.

The New Delhi Declaration is a useful initial guide. It expresses the view that sustainable development requires a comprehensive and integrated approach to economic, social and political processes. It adopts seven principles[80] that are widely accepted as relevant to sustainable development. However, not all of them are accepted as principles or rules of international law. In the following, we will briefly discuss their contents and current legal status.

(1) Sustainable use   States have the sovereign right to exploit their natural resources. This includes the right to be free from external interference over their exploitation.[81] The right is qualified by the duty of the state to not cause irreparable damage to the territories of other states. Beyond this duty, there is increasing support for requiring the state not to cause irreparable damage to the global environment. However, such duty is not yet recognized by international law.

The (sliding) scale of transboundary responsibilities is expressed in Principle 21 of the Stockholm Declaration which provides that: 'States have, in accordance with the Charter of the United Nations and principles of international law, the sovereign

---

80   [See above, 38].

81   The right to permanent sovereignty over natural resources has been consistently affirmed since its first adoption by the UN General Assembly Resolution in 1952.

right to exploit their own resources pursuant to their own environmental policies, and the responsibility to ensure that activities within their jurisdiction or control do not cause damage to the environment of other States or areas beyond the limits of national jurisdiction'. This Principle 21 was reaffirmed by Principle 2 of the Rio Declaration[82] and reflects a rule of customary law.[83]

It could be argued that the transboundary nature of ecological systems needs to translate into a basic obligation of sustainable use of natural resources.[84] A 'sustainable use' obligation is recognized in rudimentary form, for example, in the Biodiversity Convention[85] and the Climate Change Convention,[86] but only with respect to a particular subject area. Notwithstanding these and other limitations, including its vagueness, 'sustainable use' is recognized as a rule that some also consider to be part of customary law.[87]

(2) Equity and eradication of poverty   The principle of equity represents the social dimension of sustainable development. It is understood to refer to intragenerational equity (the rights of people within the current generation of fair access to the Earth's natural resources) and intergenerational equity (the rights of future generations). Intragenerational equity includes the commitment of states to eradication of poverty as, for example, expressed in the Millennium Development Goals and the Johannesburg Plan of Implementation. The 'principle' of poverty eradication is, however, a political objective without legal obligations. The same is true for the 'principle' of equity, although the Brundtland Report refers to it as of 'overriding priority'[88] for sustainable development. A legal foundation for both equity and poverty eradication can be found in various human rights treaties,[89] the Universal Declaration of Human Rights[90] and the Declaration on the Right to Development.

(3) Common but differentiated responsibilities   As an expression of the principle of co-operation[91] states have common responsibilities towards protecting the

---

82   The Rio Declaration 1992 (n. 91 above) added just two words: 'their own environmental *and developmental* policies' (my emphasis).

83   Sands 2003 (n. 1 above), 241; Beyerlin 2007 (n. 16 above), 14. For the actual extent of this rule see Chapter 5.

84   Cordonier Segger and Khalfan 2004 (n. 30 above), 111, 117–22; and Chapter 5.

85   Convention on Biological Diversity 1992 (n. 65 above), Article 10 qualifies 'sustainable use … as far as possible and as appropriate'.

86   Convention on Climate Change 1992 (n. 10 above), Article 3(4).

87   Beyerlin 2007 (n. 16 above), 20.

88   World Commission on Environment and Development (1987), *Our Common Future*, 'Brundtland Report' (Oxford and New York, Oxford University Press), 13.

89   Including the Convention on the Elimination of All Forms of Discrimination Against Women 1979 and the Convention on the Rights of the Child 1989.

90   Not including the rights of future generations.

91   See Principle 7 of the Rio Declaration on Environment and Development (the Rio Declaration) (Rio de Janeiro, 13 June 1992; UN Doc. A/Conf.151/26 (vol. I); 31 ILM 874 (1992)): 'States shall co-operate in a spirit of global partnership to conserve, protect and restore the integrity of the Earth's ecosystem'.

global environment. In view of the different contributions to global environmental degradation, however, the responsibilities are differentiated. The differentiation can also be seen as a reflection of the principle of equity. The content of the common but differentiated responsibility principle is defined in Principle 7 of the Rio Declaration: 'The developed countries acknowledge the responsibility that they bear in the international pursuit of sustainable development in view of the pressures their societies place on the global environment and of the technologies and financial resources that they command'. While the mentioned cooperation principle is recognized as international customary law, its application with respect to equity, somewhat surprisingly, is not. States have not yet accepted the idea of differentiated responsibilities as 'general practice'. With this uncertainty and the fact that it has not consistently been used in environmental treaties,[92] the principle of common but differentiated responsibilities can only be perceived as a 'nascent' principle of international law.[93]

(4) The precautionary approach to human health, natural resources and ecosystems   The precautionary approach is central to sustainable development and recognized as a norm of international customary law. However, legal normativity only applies to the general duty to take action when the state is faced with causing environmental damage if no action is taken. This approach, as opposed to a 'principle',[94] is expressed in Principle 15 of the Rio Declaration[95] and has been repeated in many hard and soft law documents. Whether such normative quality amounts to a rule[96] or, considering the indetermination of required action, to a principle, is a matter of judgement.

The indifference of the precautionary approach, both in terms of *when* and *what* action is required, poses a severe limitation on its effectiveness. This is not only so at international level, but also at national level. In Germany, for example, where the precautionary principle (*Vorsorgegebot*) originated and where it has been heralded as 'the most important principle of German environmental policy',[97] the experience has been sobering. Its impact has been visible in the policy area, as a general 'approach', but hardly in law. In international as well as national environmental law, there is a remarkable gap between the 'promise' of the precautionary principle and its actual application.

---

92   It is fully applied only in the Climate Change Convention 1992 (n. 10 above) and the Kyoto Protocol 1997.

93   Cordonier Segger and Khalfan 2004 (n. 30 above), 132; Beyerlin 2007 (n. 16 above), 17.

94   Requiring clear direction and legal consequence.

95   Rio Declaration 1992 (n. 91 above): 'Where there are threats of serious or irreversible damage, lack of full scientific certainty shall not be used as a reason for postponing cost-effective measures to prevent environmental damage'.

96   Beyerlin 2007 (n. 16 above), 15–16.

97   See Cordonier Segger and Khalfan 2004 (n. 30 above), 144 at footnotes for further references.

(5) Public participation and access to information and justice

and

(6) Good governance   Both these principles are procedural in nature and not limited to environmental decision-making. To this end, they belong to any modern day democracy. For sustainable development, however, they are essential. Without effective participation of civil society and transparency of governance, sustainable development will remain an unfulfilled promise. International law has acknowledged this at least, in all its soft law documents related to sustainable development. Agenda 21, for example, promises that 'states will ensure broad public participation in initiatives for sustainable development, through access to information and access to justice'.[98] The legal basis for these participatory rights of the public is ultimately human rights (in treaties and constitutions). On the other hand, a key aspect of achieving sustainable development, the 'right to development'[99] is not yet recognized by international law. The principle of good governance could also be traced to human rights, but is used only as a term to describe democracy, transparency, effectiveness, procedural rights and the rule of law as prerequisites for sustainable development. In this sense, the Johannesburg Declaration commits 'to undertake to strengthen and improve governance at all levels'.[100]

(7) Integration and interrelationship, in particular in relation to human rights and social, economic and environmental objectives   The principle of integration is another procedural principle relevant to sustainable development. It is often treated as the most central of all and features prominently in many declarations on sustainable development. The need for integration was reinforced in the Johannesburg Declaration, where states proclaimed 'a collective responsibility to advance and strengthen the interdependent and mutually reinforcing pillars of sustainable development – economic development, social development and environmental protection'. A 'collective responsibility' is, however, not the same as a legal requirement. The norm-creating character of integration remains in doubt as long as it lacks content and direction. For example, is mere awareness of social, economic and environmental interrelations already 'integration' or does this require a certain methodology and strategy? Further, what political, legal and institutional arrangements are required? And most importantly, what parameters and objectives

---

98   Agenda 21: Programme of Action for Sustainable Development (Agenda 21) (Rio de Janeiro, 14 June 1992; UN Doc. A/Conf.151/26 (1992) 31 ILM 874 (1992)), paras. 8.3(d), 8.4(e) and 23.2; See also Principle 10 of the Rio Declaration 1992 (n. 91 above).

99   Defined in Article 1 of the (non-binding) Declaration on the Right to Development 1986 (4 December 1986; UN Doc. A/41/53) as 'an inalienable human right by virtue of which every human person and all peoples are entitled to participate in, contribute to, and enjoy social, cultural and political development, in which all human rights and fundamental freedoms can be fully recognized'.

100   Johannesburg Declaration 2002 (n. 66 above), para. 10.

exist for an 'integrated' decision-making process? If they are not defined, what reason is there to believe that integration will lead to sustainable development?

To operationalize the principle of integration, Cordonier Segger and Khalfan suggest 'four degrees of integration' ranging from 'separate spheres', to 'parallel yet interdependent', to 'partially integrated spheres' and finally to 'highly integrated new regimes'.[101] For instance, many environmental regimes exist in parallel, yet interdependent with trade regimes. To achieve a higher degree of integration both regimes need to merge into one new regime. As an example of such a 'highly integrated' regime, Cordonier Segger and Khalfan refer to the 2000 Cartagena Protocol on Biosafety. This regime contains trade regulations as well as social and environmental regulations. The question remains, however, on what basis eventual conflicts between the various aspects should be resolved.

Overlooking these seven principles proposed by the International Law Association, we can easily see their relevance for sustainable development. They describe prerequisites and ingredients of the process. All principles are resonant of existing international environmental law and legal developments in many states although, in most cases, they are only emerging principles of international law. Their normative quality is often weak or non-existent.

For the substance of sustainable development, the principles have limited value. They are supportive of the concept, but do little to determine its specific content. Assessing their actual contents and their status in international law, we are left with the impression that they are not reaching the core idea of sustainable development. We need to spell out what this core is and define a fundamental principle that gives shape and direction to these other principles. We will examine this fundamental character now.

The Fundamental Character of Sustainability

The concept of sustainable development owes its meaning and legal status to the principle of sustainability. As discussed above, the presupposition of the sustainability principle is, in fact, the only way to give the integrative character of sustainable development direction and form. We can now see how fundamental sustainability is to the concept of sustainable development.

This characterization has three important implications for the sustainability discourse. The first is that sustainability is separate from sustainable development. Both terms are often used interchangeably, but need to be kept separated from each other. The second implication is that the notion 'sustainable development' relates development to sustainability in a sense that the former is grounded in the latter. Like 'sustainable management', 'sustainable use' and similar composite terms, 'sustainable development' represents an application of the principle of sustainability, nothing more and nothing less. The third implication is that sustainability is the most fundamental of all environmental principles, although this fundamentality has yet to be fully recognized in law and governance.

As argued from the outset of this book, there are important parallels between the idea of sustainability and the idea of justice. The justice discourse has always

---

101   Cordonier Segger and Khalfan 2004 (n. 30 above), 106–8.

maintained certain distinctions that are equally relevant to the sustainability discourse. First, justice is different from composite terms such as 'just society'. Second, the notion of a 'just society' relates society to justice in a sense that the former is grounded in the latter. Third, the term 'just society' represents an application of the principle of justice which is fundamental to civilized nations similarly to the principles of freedom, equality – and sustainability.

The characteristic of fundamental principles is that they cannot *per se* be defined in precise terms, yet they are absolutely indispensable as guiding ideals for the design of public policy. Governments often fail to live up to these ideals, but they are constitutionally obliged to pursue them. The practical value of fundamental principles is that they give us a yardstick for assessing any policy measure. It is of no relevance whether agreement on such assessment can be achieved. An open democratic society lives from the competition of ideas. It is crucial, however, that ideas can be communicated and compared. If there are no ideas emphasizing the importance of a sustainable society, then logically they will not influence the public discourse.

The prime responsibility of law is to promote fundamental principles, often expressed in constitutions and human rights catalogues, and ensure that the legal process is reflective of them. If sustainability is perceived as one of such fundamental principles, the legal process will have to be reflective of it. If the principle of sustainability is perceived as just one of an array of environmental principles, it will compete among these and almost certainly vanish in the politics of governments that are currently obsessed with economic growth and international competition.

We will now discuss to what extent the principle of sustainability has, in fact, entered the legal system. The example of three different levels of law can illustrate current achievements and shortcomings of this pursuit.

The first level is domestic law using the example of New Zealand as the country with the reputation of having the world's most advanced environmental legislation. The second chosen level is the international judiciary system. International courts and tribunals are the safeguards of international law, in charge of judging states' behaviour. They are a good indicator of the UN system's 'legal consciousness'. The third level is international law itself, i.e. the body of international rules and agreements related to sustainability.

The New Zealand experience New Zealand was the first country to apply the principle of sustainability in law. During the 1980s New Zealand's law and institutions of environmental governance underwent a radical reform. At its core was the shift from traditional environmental protection and resource management to an integrated concept of 'sustainable management'.[102] The coordinating Ministry for the Environment stated the fundamental importance of sustainability as follows:

---

102 Incorporated, for example, in the Resource Management Act 1991 and the Local Government Act 2002. See Grinlinton, D. (2002), 'Contemporary Environmental Law in New Zealand', in Bosselmann, K. and Grinlinton, D. (eds), *Environmental Law for a Sustainable Society* (Auckland, New Zealand Centre for Environmental Law), vol. 1, 19–46, 26–30.

'Sustainability is a general concept and should be applied in law in much the same way as other general concepts such as liberty, equality and justice'.[103]

This statement expressed the sentiments of many. And although it did not spark a legal revolution at the time,[104] the likening of sustainability to 'other general concepts such as liberty, equality and justice' has set a new agenda not only in New Zealand, but around the world. Similar comments have been made by the Brundtland Commission[105] and at the Earth Summit 1992 in Rio.[106] The challenge for jurisprudence and law has been to integrate sustainability as a fundamental principle.

In 1991, New Zealand passed the Resource Management Act to introduce sustainability as a core concept of decision-making. Conceptualizing it as 'sustainable management', section 5 (2) of the RMA defines as follows:

2. In this Act, 'sustainable management' means managing the use, development, and protection of natural and physical resources in a way, or at a rate, which enables people and communities to provide for their social, economic, and cultural wellbeing and for their health and safety while –

a. Sustaining the potential of natural and physical resources (excluding minerals) to meet the reasonable foreseeable needs of future generations; and

b. Safeguarding the life-supporting capacity of air, water, soil, and ecosystems; and

c. Avoiding, remedying, or mitigating any adverse effects of activities on the environment.[107]

This definition and its application has been the subject of many debates and decisions in the Environment Court. The Environment Court, by the way, is another world's first. Established in 1995, it replaced the former Planning Tribunal to provide the expertise necessary for environmental law cases under the RMA.

During its environmental jurisprudence, the Environment Court developed two different approaches to interpreting section 5 of the RMA – the so-called 'environmental bottom line approach' and the so-called 'overall judgement approach'. Two cases, _Foxley Engineering Ltd v. Wellington City Council_, Decision W12/94, and _Campbell v. Southland District Council_, Decision W114/94, cite the use of the environmental bottom line approach.[108] In _Foxley Engineering Ltd_, there was an appeal from a decision of the Wellington City Council granting resource consent to establish a service station and carpark. At the time the Court was still known as the Planning Tribunal and in its decision referred to section 5 and said, 'The provisions of s.(5)(2)(a) (b) (c) may be considered cumulative safeguards

---

103　New Zealand Ministry for the Environment (1989), 'Resource Management Law Reform: Sustainability, Intrinsic Values and the Needs of Future Generations', Working Paper No. 24 (Wellington), 9.

104　On 'system failures' of New Zealand's environmental law reform see, for example, Grinlinton 2002 (n. 103 above), 37–46.

105　Brundtland Report (n. 88 above).

106　Rio Declaration (n. 99 above).

107　Grinlinton 2002 (n. 103 above), 26.

108　Skelton, P. and Memon, A. (2002), 'Adopting Sustainability as an Overarching Environmental Policy: A Review of section 5 of the RMA', _Resource Management Journal_ 10:1, March 2002, 8–9, 6.

... They are safeguards which must be met before the Act's purpose is fulfilled. The promotion of sustainable management has to be determined therefore in the context of these qualifications which may be accorded the same legal weight'.[109] The Tribunal revoked the resource consent based on the potential adverse effects on the inner city amenities and heritage values.

In *Campbell* there was an appeal against the granting of resource consent by the Southland District Council to allow an international airport in Southland. The Tribunal repeated the view expressed above in the quote from *Foxley Engineering Ltd* and also cancelled the consent. It held that 'Section 5 is not about achieving a balance between benefits occurring from an activity and its adverse effects',[110] thus favouring a fundamental sustainability approach.

The High Court decision of *New Zealand Rail Ltd v. Marlborough District Council*[111] made some similar observations about the RMA. 'There is a deliberate openness about the language, its meanings and its connotations which I think is intended to allow the application of policy in a general and broad way'.[112] This observation is meant to convey that the purpose definition in section 5 needs to be understood as an overarching commitment to achieving sustainability.[113]

Another case, *Trio Holdings v. Marlborough District Council*,[114] involved a marine farming application to grow sponges from which anti-cancer compounds were to be extracted. The Court concluded that the adverse effects were not so serious as to be inconsistent with the requirement for sustainable management. Here, the Court performed a balancing test, weighing the benefits of the proposed venture against the possibility of environmental harm and noted that, 'the proposal has the potential to provide for the social and economic wellbeing of the communities' and had difficulty 'accepting that the visual impairment from the buoys of the sponge farm ... should prevail over an issue of the national health and welfare which stems from the implications of this proposal'.[115]

The last case to be mentioned here is the *North Shore City v. Auckland Regional Council* (Okura)[116] which involved the exclusion of the Okura catchment from the urban limits of North Shore City. Now the term 'overall broad judgment approach' made its appearance. In this case the Court set out a different statement of how the purpose is to be applied:

---

109   Ibid.

110   Upton, S., Atkins, H. and Willis, G. (2002), 'Section 5 Re-visited; A critique of Skelton and Memon's analysis', *Resource Management Journal* 10:3, November 2002, 10–22, 16.

111   *New Zealand Rail Ltd v. Marlborough District Council* [1994] NZRMA 70.

112   Upton 2002 (n. 111 above), 17.

113   Curran, S. (2004), 'Sustainable Development v Sustainable Management: The interface between the Local Government Act and the Resource Management Act', *New Zealand Journal of Environmental Law* 8, 267–94, 279.

114   *Trio Holdings v. Marlborough District Council* [1997] NZRMA 97.

115   Upton, Atkins and Willis 2002 (n. 111 above), 17.

116   *North Shore City v. Auckland Regional Council* [1997] NZRMA 59.

The method of applying s.5 then involves an overall broad judgment of whether a proposal would promote the sustainable management of natural and physical resources. That recognizes that the Act has a single purpose ... Such a judgment allows for comparison of conflicting considerations and the scale of degree of them, and their relative significance or proportion in the final outcome.[117]

This refinement of the balancing and weighing approach was supported by Memon and Skelton[118] and since the role of the Act is essentially a 'conflict resolving statute' there will be an element of weighing.

However, other commentators maintain that under a straightforward reading of section 5, resource managers need to secure the outcomes detailed in paragraphs (a), (b), and (c) which operate at high level constraints.[119] If the overall judgment approach of weighing socio-economic merits and environmental effects of a proposal is used, they perceive an undermining of the sustainability principle.

We can see from the New Zealand experience that sustainability plays an important role in the Court decisions, mainly how it is interpreted in the RMA. However, this experience also suggests that neither well-written legislation nor the existence of a specialist Environment Court could *per se* make a difference. Obviously, reasoning around the fundamental importance of sustainability has had some impact on the way judges approach environmental cases. Leading judge Peter Salmon has repeatedly stated the fundamental importance of the sustainability principle 'as the only meaningful cure to the problems that face the world'.[120] The fact that there is little consistency in the reasoning of the Courts may be explained by a too slow learning process. When the former Prime Minister Sir Geoffrey Palmer reviewed the making of the Resource Management Act, he expressed regret 'that Parliament did not abolish the Planning Tribunal when the new legislation was framed. ... The need to change the judicial culture was overlooked'.[121] In a similar vein, the Parliamentary Commissioner for the Environment (PCE) criticized authorities and Courts for not sufficiently focusing on the Act's 'core thrust' with its recognition of 'intrinsic values' and ecological 'bottom lines'.[122] The PCE has repeatedly reminded the people of New Zealand that sustainability is a foundational principle for society and its economy ('strong sustainability') requiring a profound shift of values and policies.[123]

---

117   Ibid. 94.

118   Memon and Skelton 2002 (n. 109 above), 9.

119   Upton, Atkins and Willis 2002 (n. 111 above), 13.

120   Salmon, P. (2002), 'Sustainable Development in New Zealand', Paper presented to the Auckland Branch Resource Management Law Association on 30 October 2002, 3.

121   Palmer, G. (1995), *Environment: The International Challenge* (Wellington, Victoria University Press), 170.

122   Office of the Parliamentary Commissioner for the Environment (1998), *Towards Sustainable Development. The Role of the Resource Management Act 1991* (Wellington), 7; Office of the Parliamentary Commissioner for the Environment (2007), *Sustainability Review 2007: New Zealand's Progress Towards Sustainable Development* (Wellington).

123   Office of the Parliamentary Commissioner for the Environment (2002), *Creating Our Future. Sustainable Development for New Zealand* (Wellington), 35.

Sustainability jurisprudence in the international judicial system   As discussed above, sustainable development does have established status as a principle of international law. This is visible in the evolving jurisprudence of international tribunals and courts as we have seen above.[124] However, as will be shown now, the principle of sustainability has not yet directly influenced the outcomes of international disputes.

In his analysis of international case law, John Gillroy found that sustainability 'has emerged as the core concept of the current environmental debate within international law'[125] and that sustainability is present in *obiter dicta*, illustrating its moral significance, but not in *rationes decidendi* of the decisions.[126] There is a notable gap between general references to sustainability and its actual recognition as a guiding legal principle. This anomaly exists for normative and institutional reasons.[127]

First, normatively, for a legal principle to guide international dispute resolution, it must not only be a legal principle, but a rule-generating adjudicatory norm. This has not occurred for sustainability because the 'principle' of sustainable development itself is not of a sufficiently definitive rule-creating character; it contains a number of competing and even contradictory sub-principles that dilute its normative power. Second, institutionally, the international judiciary system has been evolving as a set of parallel, closed legal regimes with specific adjudicatory norms. Therefore, a new legal principle, in order to become an adjudicatory norm, requires institutional refinements. A specialist International Environmental Court may be required, although the New Zealand experience would caution against any reliance on institutional reform.

Fundamentally, international law is shaped around the core value of state sovereignty. This has not changed over the past 60 years despite new challenges to sovereignty, for example, through the emergence of human rights as universal norms or the emergence of global concerns such as economic liberalization and environmental sustainability. The International Court of Justice (ICJ) has not altered its core adjudicatory norm of sovereignty to accommodate sustainable development. And while the International Tribunal for the Law of the Sea (ITLOS) has its focus on the law of the sea, the Panel and Appellate Body of the World Trade Organization (WTO), on free trade and the UN Committee on Human Rights (CHR), on human dignity, none of these tribunals have referred to sustainable development in any other way than *obiter dicta*. Quite obviously, the norm-generating quality of sustainable development has not been recognized.

---

124   See above, 24–5.

125   Gillroy, J.M. (2006), 'Adjudication Norms, Dispute Settlement Regimes and International Tribunals: The Status of "Environmental Sustainability" in International Jurisprudence', *Stanford Journal of International Law* 42, 1–52.

126   Ibid. 5–6.

127   Bosselmann, K. (2008), 'The Environmental Jurisprudence of International Tribunals: Making Sustainability Count', in Kotze, L. and Robinson, N. (eds), *Compliance and Enforcement. Toward More Effective Implementation of Environmental Law* (Cambridge, Cambridge University Press).

To illustrate these findings we will briefly discuss some key cases.[128]

Until the mid 1990s, the ICJ considered sustainability issues in passing remarks and not with respect to specific obligations of states. For example, in *Jan Mayen Denmark v. Norway* (1993),[129] Denmark asked the ICJ to decide where a delimitation line should be drawn between Denmark and Norway's fishing zones and continental shelf areas in the waters between Greenland and the island of Jan Mayen. In its decision, the ICJ applied the principle of equity between sovereign states rather than a sustainability related principle more akin to ecosystem boundaries. In a Separate Opinion, Judge Weeramantry expressed his agreement with the judgment of the Court, but examined the special role played by equity in the decision.[130] He noted that environmental factors were crucial in maritime boundary delimitations and should be considered an integral part of the Court's concern for equity. He connected equity to both the environment and the welfare of future generations. This notion was one of the first explicit arguments for the components of sustainable development within the Court's *dicta* reasoning.

The leading ICJ case is *Gabçikovo-Nagymaros Hungary v. Slovakia* (1997).[131] This case concerned a hydroelectric dam project to be built on the Danube River. Project planning began in 1977 after Hungary and (the former) Czechoslovakia signed a bilateral treaty. In 1989 Hungary suspended the project and by 1992 it tried to pull out of the project because it would divert 80 per cent of the flow of the Danube away from Hungary. Hungary cited ecological necessity as its basis for withdrawing from the treaty and stopping the project, and so the Court was faced with the argument of sustainability and environmental damage as well as the usual questions of law of watercourses, state responsibility, and law of treaties. The Court readily acknowledged that the concerns expressed by Hungary for its natural environment in the region affected by the project related to an 'essential interest' of that state.

While the Court held that these arguments were insufficient to terminate the 1977 treaty, it did connect sustainability concerns to international law and requested that the parties renegotiate the treaty:

> It is clear that the Project's impact upon, and its implication for, the environment are of necessity a key issue ... Through the ages, mankind has, for economic and other reasons, constantly interfered with nature. In the past, this was often done without consideration of the effects on the environment. Owing to new scientific insights and to a growing awareness of the risks for mankind – for present and future generations – new norms and

---

128   For a detailed examination see Bosselmann, K. (2008), 'The Environmental Jurisprudence of International Tribunals: Making Sustainability Count', in Kotze, L. and Robinson, N. (eds), *Compliance and Enforcement. Toward More Effective Implementation of Environmental Law* (Cambridge, Cambridge University Press), 27.

129   *Case Concerning Maritime Delimitation in the Area between Greenland and Jan Mayen (Denmark/Norway)*, ICJ 38; 99 ILR 395 (1993).

130   Ibid., Separate Opinion of Judge Weeramantry.

131   *Gabçikovo-Nagmaros Case* 1997 (n. 44 above), available at <http://www.icj-cij.org/docket/index.php?sum=483&code=hs&p1=3&p2=3&case=92&k=8d&p3=5>, accessed 1 December 2007.

standards have been developed, set forth in a great number of instruments during the last two decades. Such new norms have to be taken into consideration, and such new standards given proper weight, not only when States contemplate new activities but also when continuing activities begun in the past. This need to reconcile economic development with the protection of the environment is aptly expressed in the concept of sustainable development.[132]

In his Separate Opinion, Vice President Weeramantry addressed three questions in relation to environmental law – the principle of sustainable development in balancing the competing demands of development and environmental protection, the principle of continuing environmental impact assessment and the question of the appropriateness of the use of an *inter partes* legal principle such as issue preclusion in the resolution of issues with *erga omnes* implications.[133] His opinion stated that the right to development and the right to environmental protection are principles currently forming part of the body of international law and that they need to be reconciled with the principle of sustainable development which is a recognized principle of international law. He considered it 'a general principle of international law recognized by civilized nations' and 'an integral part of modern international law', 'by reason not only of its inescapable logical necessity, but also by reason of its wide and general acceptance by the global community'.[134]

The Court ended up ruling that the treaty remained in effect and that both parties were responsible for breeches of state responsibility. However, sustainability did play a minor role in the decision of the Court when they directed the parties how to renegotiate the treaty by including balance between environment and development and in the use of environmental impact studies to structure the new project.

Until today, the ICJ has relied on the well-established body of sovereignty-centred international law, but there has been an increased recognition of environmental boundaries. However, sustainability has not reached a status capable of shaping the *ratio decidendi* of the ICJ. Sustainability is still mainly *obiter dicta* when considered against the 'trump card' of state sovereignty.

Compared to the ICJ, the legal regime of the ITLOS appears more conducive to the concept of sustainability with a focus on the resources of the high seas, their use and their conservation for present and future generations. For example, the Preamble to the UN Convention on the Law of the Sea seeks to establish ' legal order for the seas and oceans which will ... promote ... the equitable and efficient utilization of their resources, the conservation of their living resources, and the study, protection and preservation of the marine environment'. However, so far there has not been an ITLOS decision favouring conservation duties over sovereignty rights.[135]

The 'trump card' of the WTO is efficiency of free trade, although the principle of sustainability is indirectly referred to in the Preamble of GATT: 'the Parties ...

---

132 Ibid., para. 140.

133 Ibid., Separate Opinion of Judge Weeramantry.

134 *Gabçikovo-Nagymoros Case* 1997 (n. 44 above), Separate Opinion of Judge Weeramantry.

135 For example see the *Southern Bluefin Tuna Case* (*Australia and New Zealand/ Japan*), 38 ILM 1624 (1999).

recognize that their relations in the field of trade and economic endeavour should be conducted with a view to raising the standards of living ... and expanding the production of and trade in goods and services while allowing for optimal use of the world's resources in accordance with the objective of sustainable development'.[136]

The relevant WTO cases all concerned the interpretation of Article XX (b) and (g) of the GATT. Article XX sets out 'General Exceptions' for free trading:

> Subject to the requirement that such measures are not applied in a manner that would constitute a means of arbitrary or unjustifiable discrimination ... nothing in this Agreement shall be construed to prevent the adoption or enforcement by any contracting party of measures:
>
> ...
>
> (b) necessary to protect human, animal or plant life and health;
>
> ...
>
> (g) relating to the conservation of exhaustible natural resources if such measures are made effective in conjunction with restrictions on domestic production or consumption.[137]

In *Tuna–Dolphin I Mexico v. US* (1991), the US government prohibited imports of yellowfin tuna from Mexico based on the Marine Mammal Protection Act (MMPA) 1972. Dolphins had a habit of running with the tuna and under the MMPA, the US required nations to create conservation fishing programmes that would reduce incidental taking of dolphins when fishing for tuna or suffer prohibitions on their importations of tuna to the US.[138] Mexico claimed discrimination against their products in trade without a just cause.[139] The US justified its direct embargo against Mexico under GATT Articles XX (b) and XX (g) which provided a general exception from other GATT obligations for measures 'necessary to protect human, animal or plant life or health' and 'relating to the conservation of exhaustible natural resources'.[140] The US argued the dolphins were an exhaustible resource and that no other means to implement MMPA requirements were available.

The GATT Panel set the burden of proof 'on the party invoking Article XX to justify its invocation'.[141] It also set a high threshold for its invocation, 'Article XX was only intended to allow contracting parties to impose trade restrictive measures to the extent that such inconsistencies were unavoidable'.[142]

---

136   Robb, C. (ed.) (2001), *International Environmental Law Reports: Trade and Environment* (Cambridge, Cambridge University Press), vol. 2, 319 para.129.

137   Ibid., 313 para. 113.

138   Gillroy 2005 (n. 126 above), 52.

139   United States Restrictions on Imports of Tuna (*Mexico v. US*) GATT Panel Report 3 September 1991 and United States Restrictions on Imports of Tuna (*European Comm. and Netherlands v. US*) GATT Panel Report 16 June 1994, both reported in 2 IELR 48–114.

140   Ibid. 52.

141   Ibid., 71 para 5.22.

142   Ibid., 72 para. 5.27.

This decision, subsequent WTO decisions such as *Tuna–Dolphin II EU Comm. and Netherlands v. US* (1994), *Turtle–Shrimps India et al. v. US* (1998) and others[143] all held that sustainability concerns are subordinated to ensuring free trade. The WTO regime clearly holds an economic bias against ecological considerations.

When we look at the international judicial regime of protecting human rights, we can detect increasing concerns for the environmental foundations of human rights.[144] The UN Human Rights Committee and the European Court of Human Rights have, in a number of cases, asserted that the enjoyment of human rights depends on a certain standard of environmental conditions. But sustainability does not feature as a component of human rights themselves.

The categorical difference between human rights and sustainability is particularly evident in cases involving indigenous human rights. Typically, indigenous peoples perceive their cultural and natural environment as part of their identity, requiring protection of both individual and collective rights. Such wider perceptions are alien to (Western) human rights ideology. For example, in *Ominayak v. Canada* (1984), the leader of the Lubicon Lake Band of the Cree Indian Band of Alberta, Canada, submitted a communication to the UN Human Rights Committee in which he argued that the Cree's land had been expropriated for oil and gas exploration and destroyed. He alleged violations by the government of Canada of the Band's right of self-determination and the right to freely dispose of its natural wealth and resources and not be deprived of its own means of subsistence contrary to Articles 1 to 3 of the 1966 International Covenant on Civil and Political Rights.[145] Canada argued that the author of the communication could not invoke Article 1 because it dealt with individual rights, but the Committee decided it could examine the case under other Articles of the Covenant including Article 27.

The Committee held that the communication violated Article 27 and 'the rights protected by Article 27 included the rights of persons, in community with others, to engage in economic and social activities which are part of the culture of the community to which they belong'.[146] However, the Committee rejected the Cree's argument that the ecology of the land was their right rather than any property and development rights. In a Separate Opinion Mr Ando disagreed with the finding of a violation of Article 27: 'It is not impossible that a certain culture is closely linked to a particular way of life and that industrial exploration of natural resources may affect the band's way of life'.[147]

We will later discuss to what extent ecological dimensions can and should define the content and scope of human rights. The current jurisprudence of international

---

143 Bosselmann 2008 (n. 128 above).

144 See below Chapter 4.

145 *Ominayak and the Lubicon Lake Band v. Canada*, UN Human Rights Committee, (communication no. 167 (1984); UN Doc. CCPR/C/38/D/167/1984) as reported in Robb, C. (ed.) (2001), *International Environmental Law Reports: Trade and Environment* (Cambridge, Cambridge University Press), vol. 3, 27.

146 Ibid. 59 para. 32.2.

147 Ibid. 60 (Separate Opinion of Mr Ando). Further cases are examined in Bosselmann 2008 (n. 128 above).

courts does not follow such an integrative approach. Rather, it sets forth the traditional separation of human rights reasoning and sustainability concerns.

Where does this leave us with our argument that sustainability is a fundamental principle of law? Domestic and international courts have increasingly resonated with sustainability concerns; however, such 'greening' of the judicial system is still far from recognizing sustainability as a fundamental principle of law.

The courts' reluctance is reflective of international law in general. As will now be shown, there has been a trend – in soft law as well as in treaty law – towards recognizing the importance of sustainable development. However, the lack of defined content has been, and continues to be, a major drawback. From all we have argued so far, it is crucial to bring the debate back to the ethical basics of sustainable development. Unless the legal discourse on sustainability includes the ethical dimension, we will not see any change. Or to put it positively, realizing the normative, i.e. ethical and jurisprudential, qualities of the sustainability principle will lead to the legal recognition and operability of the sustainable development concept. The following discussion of recent international agreements can help us to further identify and appreciate these qualities.

The Changing Architecture of International Decision-making

In Chapter 1 we have seen that a number of documents preceding the 1987 Brundtland Report have recognized the fundamental importance of sustainability. Among these were the 1972 Stockholm Declaration, the 1980 World Conservation Strategy and the 1983 World Charter for Nature.[148] But also the Brundtland Report itself and various subsequent documents acknowledged the priority of ecological sustainability. Among these were the 1991 document Caring for the Earth and the 1996 IUCN Draft Covenant for Environment and Development. Even the 1992 Rio Declaration and Agenda 21 reflected the notion that ecological sustainability is indispensable for social and economic development.[149] These and other state-negotiated agreements 'only' lack the definitive clarity that is needed to make a real difference.

Maurice Strong, the Secretary General of the Earth Summit, commented:

> It became evident that world leaders at the Earth Summit in Rio de Janeiro in 1992 were not sufficiently compelled by the motivation of their people to accept my proposal, building on the recommendation of the Brundtland Commission and extensive consultation with others, to begin negotiation of an Earth Charter as a statement of principles designed to guide the behaviour of people and nations towards the Earth and each other.[150]

At the Rio Global Forum NGO delegates from 19 countries drafted an early version of an Earth Charter. To complete the unfinished business of the Rio Declaration, the NGO Earth Charter postulated the new ethics of sustainability by placing the

---

148   [See above, 28–29].

149   [See above, 35–37].

150   Strong, M. (2005), 'A People's Earth Charter', in Blaze Corcoran, P. et al. (eds), *Toward a Sustainable World: The Earth Charter in Action* (The Hague, Kluwer International), 11.

Earth, not just humans, into the centre of concern.[151] This approach has profound ramifications for global justice, for human rights and obligations, indeed for the entire system of law and governance.

Comparing the NGO Earth Charter with the Rio Declaration, Prue Taylor describes, in great detail,[152] the lack of Earth focus and interlinks between sustainability, wealth/poverty, war/peace, consumption, etc. as the decisive flaw of the Rio Declaration. The Declaration merely lists relevant issues without making the links and providing direction. For example, while Principle 12 of the Declaration states that 'trade policy measures for environmental purposes should not constitute a means of arbitrary or unjustifiable discrimination', Principle 4 of the Charter says: 'Trade practices and transnational corporations must not cause environmental degradation and should be controlled in order to achieve social justice, equitable trade and solidarity with ecological principles'. The former promotes trade for its own sake, the latter considers trade as means to achieve sustainability. Principles 24 and 25 of the Declaration are concerned with the conduct of war in accordance with international law and environmental protection, while Principle 5 of the Charter promotes peace as a means to eradicate poverty and achieve social justice and ecological well-being. Principle 2 of the Declaration reaffirms the 'sovereign right [of states] to exploit their own resources pursuant to their own environmental and developmental policies', while Principle 4 of the Charter states: 'We recognize that national barriers do not generally conform to Earth's ecological realities. National sovereignty does not mean sanctuary from our collective responsibility to protect and restore Earth's ecosystems'. According to Taylor, the former cements and immunizes state sovereignty, the latter subjugates state sovereignty in favour of collective responsibility.[153] Finally, the Declaration is totally focused on self-interests of people and nations, while the Charter accepts 'a shared responsibility to protect and restore Earth and to allow wise and equitable use of resources as to achieve ecological balance and new social, economic and spiritual values' (Preamble). The entire text of the Charter is concerned with the right balance between entitlements and obligations within a finite Earth.

Rio not only saw the difference between genuine ('strong') sustainability[154] and weak sustainability, it also saw the emergence of civil society as the true driver of the sustainability agenda. Clearly, states have relinquished their political leadership, but this does not mean that the sustainability discourse has vanished over the years. It only means that the discourse is led by civil society (movements, activists, academics) with states not really taking part in it. The Earth Charter process, starting in 1992 and growing ever since, is the political manifestation of the fact that sustainability

---

151   For example, 'We are Earth, the people, plants and animals, rains and oceans, breath of the forest and flow of the sea' (Preamble) or 'We agree to respect, encourage, protect and restore Earth's ecosystems to ensure biological and cultural diversity'(Principle 1) as contained in Appendix F in Taylor, P. (1998), *An Ecological Approach to International Law* (London and New York, Routledge), 379–81.

152   Ibid. 323–48.

153   Ibid. 334.

154   In addition to the Earth Charter, the NGO Global Forum negotiated 45 so-called alternative treaties reflecting a fundamental commitment to sustainability.

is neither forgotten nor 'overridden' by globalization. Only from a strictly state-centred perspective could it be argued that sustainability has lost its fundamental significance.

The Earth Charter process has led to great clarity as to where international law needs to go. As shown in Chapter 1, the process towards the Charter's adoption in 2000 was more inclusive than any treaty negotiations and represents a broader global consensus than any other international document. Furthermore, the recognition that the Earth Charter has found at the 2002 Johannesburg World Summit, by UNESCO, the IUCN and numerous local, regional and national governments is a clear indication of a changed architecture in international decision-making. To this end, civil society has *already* made inroads into international law.

From the perspective of international law, the Earth Charter is a very innovative and remarkable instrument[155] with its authority rooted in several important factors. First, because of the worldwide, cross-cultural, inter-religious, interdisciplinary dialogue that produced the Charter, and because of its extraordinary breadth and the scope of expertise involved, the drafting process succeeded where other international declaration drafting processes have not, namely in integrating the insights of science, ethics, religion, international law, and indigenous peoples. Second, the Earth Charter is, in part, grounded in established international law. The entire text endeavours to articulate and integrate the values and principles that the United Nations, international law, and the emerging global civil society have identified as widely shared and essential. Each of the principles builds upon, interconnects, and expands the ethical vision found in other international declarations and covenants. Further, the Earth Charter incorporates treaty law such as the UN Climate Change Convention, the UN Convention on Biological Diversity, and the UN Convention on Desertification.

The third factor relates to the ongoing endorsement process, which has involved over 2,500 organizations and institutions including global bodies, states, regional authorities and cities.[156] In addition, the Earth Charter is the key reference document in the current UN Decade of Education for Sustainable Development (2005–2014).[157] Millions of students in schools, universities and other educational institutions are now learning about the Earth Charter and its principles. The Earth Charter also enjoys considerable recognition in legal education, for example, in environmental law courses.[158]

A fourth factor is the increased recognition of the Charter in the theory of international law. Leading texts of international law and numerous research papers

---

155 Bosselmann, K. and Taylor, P. (2005), 'The Significance of the Earth Charter in International Law', in Blaze Corcoran, P. et al. (eds), *Toward a Sustainable World: The Earth Charter in Action* (The Hague, Kluwer International).

156 For the current state of endorsements see <http://www.earthcharterinaction.org/about_charter.html>, accessed 1 December 2007.

157 <http://www.unesco.org> (home page), accessed 1 December 2007.

158 Taylor, P. and Bosselmann, K. (2007), 'The Earth Charter in the Classroom: Transforming the Role of Law', in *Education for Sustainable Development in Action – Good Practices using the Earth Charter* (San José, Costa Rica, UNESCO and Earth Charter International), 147–50.

have examined its legal significance.[159] Recognition among international law scholars counts for a subsidiary source of international law (Article 59 of the Statute of the International Court of Justice). While the legal status of a number of the Earth Charter's principles continues to be in dispute, most of them are frequently referred to in treaties, conventions and other binding documents.

All these four factors prove the Charter's significance as a document of international law, but equally, and more importantly, they validate its message. As a declaration of principles for a just, sustainable, and peaceful world, the Charter reflects the fundamental importance of sustainability as an ethical and law-generating principle.

The key to understanding this fundamental importance is the first of the Earth Charter's four main themes entitled 'I. Respect and Care for the Community of Life' (i.e. Principles 1 to 4).

Principle 1 ('Respect Earth and life in all its diversity') explains this main theme to mean:

a. Recognize that all beings are interdependent and every form of life has value regardless of its worth to human beings.
b. Affirm faith in the inherent dignity of all human beings and the intellectual, artistic, ethical, and spiritual potential of human beings.

This short statement sums up what sustainability is all about: the recognition of the inherent value of all life and faith in the dignity and potential of human beings. Effectively, all following 15 operative principles of the Earth Charter are explanatory notes on this succinct description of ecological and human dignity.

Ecological integrity is defined as the second main theme 'II. Ecological Integrity' (with Principles 5 to 8).

Human dignity and potential are defined in the third main theme 'Social and Economic Justice' (with Principles 9 to 12) and the fourth main theme 'Democracy, Nonviolence, and Peace' (with Principles 13 to 16).

We can also say that the preservation of ecological integrity is the end of the sustainability principle and the integrity and potential of human beings the means to get there. The former describes the substance, the latter the process of sustainability.

One of the key messages of the Earth Charter is to not assume any rivalry between the natural and the human spheres. Human beings are part of nature and, although distinct through cultural arrangements, are not in an ecological sense different from it. Ecological sustainability (of all life), therefore, must not be understood as in any way competing with social and economic prosperity. It is simply the basis of both. The subject matter of ecological sustainability is described by the notion of 'ecological integrity' (second Main Principle). Apart from the terminological beauty of combining natural integrity with human integrity (expressed as 'dignity'

---

159   Kiss and Shelton 2004 (n. 14 above), 70; Taylor, P. (1999), 'From Environmental to Ecological Human Rights: A New Dynamic in International Law?', *Georgetown International Environmental Law Review* 10, 309–97; Taylor 1998 (n. 152 above), 326.

and 'potential'), ecological integrity is a scientific and ethical concept. Defined in four principles,[160] the protection and restoration of ecological integrity requires both best available knowledge about ecological facts and ethical commitment to its overarching importance. Humans do not compete with life on Earth, but aim to exist as an integral part of it. Again, this will not be possible without sound knowledge and sound morality.

The concern for the Earth's ecological systems is at the core of the sustainability principle. No other meaning is possible as the reference to ecological systems and their integrity. Through the Earth Charter we perhaps now have the 'missing link' that stood between those who favour sustainable development without being serious about it and those who would be serious about it, but were deterred by those not taking it seriously. If it is possible to perceive the integrity of ecosystems as a common concern of humanity,[161] then we have captured the essence of sustainability. There may be, and should be, ongoing debate about the means of preserving ecological integrity, but the end itself should not be disputed. If we do, we shall engage in a literally life-threatening exercise.

As we accept the protection of ecological integrity as the core meaning of the sustainability principle, we realize both its clarity and fundamentality. Sustainability aims to preserve the (measurable) integrity of ecosystems while, at the same time, acknowledging that humans are part of these ecosystems. In pursuing the protection of ecological integrity, sustainability reflects the most basic concern of human existence, namely the desire to live, survive and reproduce.

Conclusion

It would be too presumptuous to think that a fundamental concern such as the one just described has guided the legislators of the New Zealand Resource Management Act, the judges of national or international tribunals, the drafters of the World Charter for Nature, the negotiators of the Rio Declaration, or the creators of the Earth Charter. More likely, there was no such conscious and coherent effort behind these various pursuits.

However, it would be even more presumptuous to assume that the mentioned activities were guided by an attempt to balance economic, social and environmental concerns. Surely, such a balancing act would not reflect what most feel when we think of the global ecological crisis. This crisis came about because of a profound

---

160   Earth Charter 2000, Principle 5 ('Protect and restore the integrity of Earth's ecological systems, with special concern for biological diversity'), Principle 6 ('Prevent harm as the best method of environmental protection and, when knowledge is limited, apply a precautionary approach'), Principle 7 ('Adopt patterns of production, consumption, and reproduction that safeguard Earth's regenerative capacities, human rights, and community well-being') and Principle 8 ('Advance the study of ecological sustainability and promote the open exchange and wide application of the knowledge acquired').

161   In a moral sense and not as a legal principle given its current limitations (e.g. climate change as a 'common concern of mankind' under the Preamble of the Convention on Climate Change 1992 (n. 10 above).

imbalance of economic, social and environmental dimensions of human activity and not as a technological glitch. The more appropriate assumption is, therefore, a fairly common acceptance that the ecological basis of human survival is at risk. If, for example, climate change is threatening our life conditions, then any trade-offs and compromises between economic prosperity and ecological sustainability are hard to justify. Today's concerns are either those for ecological sustainability or do not exist at all (favouring a business-as-usual or overly naïve approach to facing the future).

In the remaining chapters of this book, we will explore the legal implications of the sustainability principle. Building upon the clarity and fundamentality of sustainability, we can ask what this means for traditional concepts of law and governance. If it is true that humanity faces unprecedented challenges to its continued survival, then we cannot rely on the validity of traditional concepts. New challenges require new responses.

What then does the principle of sustainability mean for the legal concept of justice? In what way may it impact on the legal protection of individual freedom and liberty? And how will the institutions of (global or national) governance have to respond to meet the challenge of ecological sustainability?

Chapter 3

Ecological Justice[1]

Introduction

The principle of sustainability aims to protect ecological systems and their integrity. Its subject matter is ecological processes. However, social processes determine to what extent and how ecological systems should be sustained. This way sustainability becomes a social issue. As there are choices to be made between competing needs and wants, questions of distributive justice arise. How sustainability affects the idea of justice will be explored in this chapter. The argument is that conventional theories of justice have been insufficient to conceptualize the environmental dimension of justice. Following recent ethical advancements, the concept of ecological justice will be developed.

The notion of ecological justice is fairly new[2] and not widely known among lawyers. This is in stark contrast to environmental justice, one of the most commonly used terms in the general environmental debate.[3] What is the difference between these aspects of justice and why is ecological justice important for law and governance?

Generally understood conceptions of 'justice' are concerned with fair distribution of social goods and burdens. Distributional concerns are in the centre of most theories of justice. They are also in the centre of justice theories with respect to environmental protection. However, there are two different relational aspects to be considered here: the justice of the distribution of the environment among people, and the justice of the relationship between humans and the rest of the natural world.

The former is usually captured by the term 'environmental justice'. What about the latter? Whether or not the relationship between humans and the rest of the natural world has anything to do with 'justice' is controversial. Those who agree speak of 'ecological justice' as a concept to include both relational aspects. Those who disagree insist that our relationship with the rest of the natural world is a matter of ethics and morality, but not justice. For them, 'environmental justice' is the only

---

1   An earlier version of this chapter appeared as Bosselmann, K. (2006), 'Ecological Justice and Law', in Richardson, B. and Wood, S. (eds), *Environmental Law for Sustainability: A Critical Reader* (Oxford, Hart Publishers), 129–63.

2   First used by Low, N. and Gleeson, B. (1998), *Justice, Society and Nature* (London, Routledge), 2; and Bosselmann, K. (1999), 'Justice and the Environment: Building Blocks for a Theory on Ecological Justice', in Bosselmann, K. and Richardson, B. (eds), *Environmental Justice and Market Mechanisms: Key Challenges for Environmental Law and Policy* (London, Kluwer Law International), 30–57.

3   Google search reveals more than a million references for environmental justice compared to 50,000 for ecological justice.

term available to describe distributional problems with respect to environmental protection.

Traditionally, legal relationships – including those of justice – are perceived as relationships purely between people: people have no legal obligations towards nature, and nature has no rights towards people. Environmental law has been moulded in this anthropocentric perception of law. However, since the early 1970s this anthropocentric perception has been challenged from an environmental ethical point of view.[4] The dichotomy between anthropocentric and ecocentric positions continues to influence the concepts of environmental law.[5]

More recently, environmental ethics has taken an interest in theories of justice. As mentioned – and in line with the general environmental law debate - there are two broad approaches. One is to keep ethics and justice strictly separate, the other is to reconceptualize justice in the light of environmental ethics. While each approach can be associated with either an anthropocentric or ecocentric perspective, it is important to note that the eco-justice debate does not entirely follow this divide. There is a considerable attempt to accommodate some form of ecocentrism within the Rawlsian conception of justice. One of the key issues in the current eco-justice debate is whether our responsibilities towards the future and the rest of the natural world can be fully addressed by extending the liberal idea of justice.

We will first see how liberal positions and ecological positions have tried to accommodate environmental ethics in the idea of justice. There is some common ground between them but liberal extensionism cannot explain why individual freedoms such as property rights should be restrained for ecological reasons. Rather than adding duties to rights, ecological realities urge us to consider redefinitions of rights themselves.

For a theory of eco-justice it is not enough to call upon environmental ethics. Unless we see all the various conceptions of justice as reflections of ethics, we will not be able to transform them. But even if we agree that 'all claims of justice are rooted in certain values other than justice itself',[6] the problem of reaching sufficient commonality of values remains. Can a commonality of values simply be assumed or would it merely be the result of a public discourse? This question leads us to the problem of how ethical norms can be institutionalized in ways that are consistent with democratic process. As will be argued, ecocentrism can neither be imposed nor will it naturally emerge from a discoursive democracy. Ecocentrism needs to be reasoned to make sense, although making sense does not necessarily emerge from reasoning in a pretext of liberal democracy.

A decisive clue in favour of eco-justice comes from the principle of sustainability. Despite the ongoing confusion surrounding the meaning of 'sustainable development',

---

4   Stone, C.D. (1974), *Should Trees Have Standing?: Toward Legal Rights for Natural Objects* (Los Altos, Kaufmann).

5   Bosselmann, K. and Grinlinton, D. (eds) (2002), *Environmental Law for a Sustainable Society* (Auckland, New Zealand Centre for Environmental Law), vol. 1, viii; Kiss, A. and Shelton, D. (2004), *International Environmental Law*, 2nd edn (New York, Transnational Publishers), 14–16.

6   Heller, A. (1987), *Beyond Justice* (Oxford, Basil Blackwell), 120.

the previous chapters have shown evidence for sustainability as an essentially ethical concept around ecocentrism. Ecological sustainability refers to the intrinsic values of 'non-human others' that can be expressed in legal concepts, not least in the idea of justice.

Eco-justice best reflects the ethics of sustainable development as we will see. Its key elements can also help to define the law related to sustainable development. Eco-justice ideas are visible in recent developments of environmental law and are likely to inform future sustainability law.

Ethical Approaches to Ecological Justice

The goal of extending justice to respond to environmental needs has long occupied theorists. In constructing a notion of eco-justice, much of the debate has centred around the distinction between the construction of a framework of basic constitutional principles which might be seen as neutral, allowing all reasonable people to agree to them, and competing notions of the individual's conception of the substantive 'good' drawn from moral, religious or philosophical doctrines that may be argued out in the democratic process. The former were termed by Rawls to be 'public reason arguments', the latter, 'comprehensive ideals'. Developments in the area of eco-justice can broadly be considered under two heads; the liberal approach and the ecological. Liberals tend to be satisfied with liberalism's defining of environmental issues as comprehensive ideals that may compete democratically. Ecologists attempt to introduce non-humans into the community of justice, the body charged with the construction of society's institutional and constitutional framework, thereby guaranteeing the consideration of nature's concerns.

The Liberal Approach

Democratic liberals attempt to extend a liberal theory of justice to include environmental concerns. They reject the contention that ecologism and democratic liberalism are incompatible. The central concern of the liberals is to reconcile liberalism's preoccupation with the individual, self-interest and state neutrality with a commitment to good environmental practice. The aim is to further environmental goals through the paradigm of liberalism in preference to a replacement. This tends to result in contingently green policy, reliant on democratic majorities, rather than a strong state commitment to ecologism.

Derek Bell focuses upon the Rawlsian distinction between public reason arguments and the separate concept of comprehensive ideals, seeking to demonstrate that environmentally friendly policies may be justified under both heads. He notes that a weak form of sustainability and a conception of environmental justice may be justified by two of the public reason arguments Rawls identifies; life sustaining properties and human health.[7] More, however, can be achieved through comprehensive

---

7   Bell, D. (2002), 'How Can Political Liberals be Environmentalists?', *Political Studies* 50, 703–24, 707–8.

arguments where majority support can be found. Thus according to Bell, Rawls's approach permits adherence to ecocentric arguments where constitutional essentials and questions of basic justice are not at issue.[8]

Bell mounts a defence of this 'environmentally friendly' liberalism by responding to objections that Rawls's theory precludes a liberal state from adhering to environmental policies even under comprehensive ideals. The objections relate to Rawls's 'difference' principle and commitment to neutrality. In relation to the former it has been argued that the difference principle, as a pre-eminent political value, requires that any action taken by the state benefits the worst-off in society. This would rule out environmental policies in circumstances where they are not of direct benefit to the impecunious in terms of income and wealth enhancement. In Bell's opinion this is inaccurate; the principle is concerned not with growth but with the distribution of income between social groups. A fair distribution of resources is more important than overall wealth maximization.[9] Because distribution is not offended by state commitment to environmental goods, neither is Rawls's difference principle.

The neutrality objection claims that the advancement of comprehensive arguments through democracy conflicts with Rawls's commitment to neutrality. It is argued that the market is a more impartial regulator than a democracy, where a minority may have their options curtailed by the majority.[10] In response Bell argues that market regulation is not fundamentally neutral, unlike democratic liberalism. Holding otherwise fails to recognize that democratic liberalism gives citizens partial control over society's common resources whereas the market cannot. A market-based approach erroneously assumes that all resources are from personal income and private wealth, ignoring the existence of collectively owned resources.[11] A market approach can only be defended through a liberal individualistic comprehensive ideal and therefore, unlike democratic liberalism, is not fundamentally neutral.[12]

Bell's thesis illustrates that liberalism and environmentalism are not incompatible. It is not clear though that beyond basic environmental goods, provided through public reason arguments, a liberal state in practice will embrace environmentalism. As Bell notes, Rawlsian liberalism is a 'contingently' rather than an 'intrinsically' green liberalism. The adoption of green policies by the state is dependent on majority support.[13]

John Barry, like Bell, has considered the compatibility of a state's embracing environmental goods with Rawls's commitment to neutrality. Barry separates out classical and social liberalism, noting that whilst the former is inconsistent with a sustainable society, the latter is not. He writes that, provided the political institutions and values of a liberal democracy can be divorced from the capitalist market-based

---

8   Ibid. 707.
9   Ibid. 715–16.
10 Ibid. 716.
11 Ibid. 719.
12 Ibid. 719–20.
13 Ibid. 721.

economy, a greened liberalism is feasible.[14] In his view social liberalism contains many of the requisite tools for a commitment to environmentalism.[15] Specifically, Barry contemplates a role for the precautionary principle in relation to irreplaceable natural resources, providing environmental protection in the face of scientific uncertainty.[16] Employment of the precautionary principle is, he asserts, consistent with Rawls's insistence on neutrality. Claims to the contrary reflect too narrow a view of neutrality, ignoring the way environmental protection can enhance liberty through protecting options.[17] He argues that because his conception of the precautionary principle does not prescribe avenues of action but rather rules out certain decisions and uses, the requirement of neutrality is not offended.[18] Furthermore, state intervention in this context rarely relates to private individuals but instead is aimed at economic actors and industrial sectors.[19]

The harm principle is also central to Barry's exposition. He sees it as redefining the notion of private property in land.[20] This involves recognition that the impact of individuals' decisions relating to their private property can impact in such a way that those decisions cannot be considered to be of a private nature.[21] The growth of regulations relating to land use is seen as an example of the state's perception of land as common, not private, property.[22] Barry though, is careful to emphasize that strict justification is required for the invocation of state intervention to avoid impinging upon the liberty of individuals.[23] In his perception both the harm and the precautionary principle are of limited strength, imposing only negative obligations not to undertake unsustainable development rather than positive prescriptions of how to act.[24] The tools Barry identifies as offered by liberalism do little more than place loose limits on the destructiveness of 'progress'.

Marcel Wissenburg also identifies tools that might serve to make liberalism capable of responding to environmental needs. He focuses his argument for a greener democratic liberalism on Rawls's 'savings principle'. The savings principle developed by Rawls aims at creating intergenerational justice through an acknowledgement by those in the original position that it is in their mutual interest to require the protection of primary goods for the next generation. Wissenburg seeks to demonstrate that the savings principle is applicable to all liberal theories not just Rawlsian or contractarian

---

14   Barry, J. (2001), 'Greening Liberal Democracy: Practice, Theory and Political Economy', in Barry, J. and Wissenburg, M. (eds), *Sustaining Liberal Democracy: Ecological Challenges and Opportunities* (Basingstoke, Palgrave), 67.

15   Ibid.
16   Ibid.
17   Ibid. 68.
18   Ibid. 71.
19   Ibid. 68.
20   Ibid. 72.
21   Ibid. 75.
22   Ibid. 74.
23   Ibid. 78.
24   Ibid. 78.

ones.[25] This is accomplished, in his view, firstly by recognizing that generations do not exist in linear formation but contemporaneously.[26] At any one time up to six generations may be interacting. Second, Wissenburg notes a new condition added by Rawls in *Political Liberalism* (Rawls 1993) to those in the original position; all generations must have followed a similar principle aimed at 'savings'.[27] In light of these two conditions, Wissenburg asserts that the savings principle must be accepted as the only rational option because it leads to mutual advantage, is in self-interest and supports individual liberty. The savings principle is to the advantage of every next generation, which makes its rejection irrational; therefore it is a necessary condition for a just society. Conversely, if the savings principle were absent from a given society that would render the society unjust as there would be a bias in favour of previous generations.[28]

Wissenburg then expands the conditions implicit in Rawls's original position that are needed to reach an agreement around the savings principle. In doing so he moves beyond the veil of ignorance, demonstrating that a savings principle is applicable to all liberal theories because it appeals to the bedrocks of liberalism; mutual advantage, self-interest and individual liberties.[29] In this way the savings principle is presented as a precursor to the existence of a just, liberal society.

Having demonstrated the necessity of the 'savings principle' to democratic liberalism, Wissenberg considers what form the 'savings principle' might take in practice. Elaborating upon the savings principle, he develops a principle of 'restraint'. The principle of restraint involves a commitment to avoid destroying goods in cases where they are irreplaceable.[30] Such a stance is defended through appeal to liberal aims of benefit maximization; options are better left open on the grounds that the present generation cannot foresee whether a good might be put to better use at a future time.[31] In situations where necessity demands the destruction of a good and an equivalent is unavailable, 'proper' compensation should be paid.[32] The effect is that each generation, as far as practicable, attempts to leave the world no worse than it was on entry.

Acceptance of the restraint principle in Wissenburg's view would have significant implications for liberalism. The principle would operate to challenge the absoluteness of private property rights.[33] There would be a shift in the burden of proof to those destroying elements of the natural world to justify their destruction

---

    25   Wissenburg, M. (1999), 'An Extension of the Rawlsian Savings Principle to Liberal Theories of Justice in General', in Dobson, A. (ed.), *Fairness and Futurity: Essays on Environmental Sustainability and Environmental Justice* (Oxford, Oxford University Press), 173–98, 174.
    26   Ibid. 177.
    27   Ibid. 175–76.
    28   Ibid. 180–81.
    29   Ibid. 183–90.
    30   Ibid. 193.
    31   Ibid. 194.
    32   Ibid. 195.
    33   Ibid. 197.

and provide a replacement or compensation.[34] Wissenburg contends that a principle of restraint would solve many of the debates around the conflict between intra- and intergenerational justice by ensuring that all attempts are made to protect the environment without sacrificing the needs of the present generation.[35] Finally Wissenburg sees the principle as responding to concerns relating to non-human nature because the principle protects the natural world insofar as where it is avoidable, no ecological being should be damaged.[36]

The Ecological Approach

The concern of ecological ethics is to draw the non-human world into the community of justice so that it is not necessary to rely wholly on democratic majorities for environmental protection. In doing so, all are careful to note that recognition of the moral value of the natural world does not indicate moral equivalence with humanity.

Nicholas Low and Brendan Gleeson argue that limiting morality to the human species is no longer defensible in light of scientific developments and the growing awareness of the interdependence between humankind and nature.[37] In response they posit two overarching principles of ecological justice to guide decision-making. First, 'ecological justice is that every natural entity is entitled to enjoy the fullness of its own form of life'. Second, 'all life forms are mutually dependent and dependent on non-life forms'.[38] Noting that the principles may operate in practice to create conflict in decision-making, they are qualified by three distinctions.[39] The first is that life has moral precedence over non-life, the second is that individualized life forms take moral precedence over life forms that only exist as communities and finally, humans take precedence over other life forms.[40] These conditions are aimed at resolving competing values and ensuring that a moral distinction between human and non-human life remains. Accepting these moral rules of thumb, Low and Gleeson consider the institutional and constitutional processes for putting ecological justice into place. Rejecting the present system as incapable of delivering ecological justice in this form they consider new models of governance.[41]

It is important to note that Low and Gleeson do not view justice as absolute but dialectic. Thus there are no universal principles, but an evolving debate about the bases of justice.[42] Conflicts are conceived of as a matter for human moral judgement, resolved by the institutions developed for such a purpose.[43]

---

34   Ibid. 198.
35   Ibid. 197.
36   Ibid. 197.
37   Low and Gleeson 1998 (n. 2 above), 155.
38   Ibid. 156.
39   Ibid. 156–7.
40   Ibid. 157.
41   Ibid. 189–93.
42   Ibid. 197.
43   Ibid. 200.

In preference to requiring ecologists to develop their own theory of distributive justice, Brian Baxter attempts to extend Brian Barry's theory of distributive justice, justice as impartiality,[44] to meet ecological concerns. This approach is chosen on the grounds that it is the most defensible of distributive justice theories.[45] Barry takes his starting point from Scanlon's modification of Rawls's original position. Scanlon's position operates not by mutual consensus but by the veto of members of the community of justice. In his conception a liberal democratic framework is among the basic constitutional rules reasonable people will not object to. Through this democratic framework conceptions of the 'good' compete for acceptance. Where a majority accepts ecological justice as a 'good' worthy of support the state may legitimately apply it. However, conversely this means that were a majority to support the extermination of whole species this would not constitute an injustice.[46] Baxter finds such a position untenable and attempts to extend Barry's theory to ensure that such an outcome is not possible.[47] He achieves this through an expansion of those involved in the formulation of the basic structure of impartial justice to include non-humans. Baxter contends that Barry has largely neglected the issue of who should comprise the community of justice, incorrectly defining its members. Barry describes his theory as solving the concerns of 'how people with competing conceptions of the good may be brought freely to live on terms which all can accept as reasonable'.[48] This conception mandates that those belonging to the community of justice must have competing conceptions of the good, thereby ruling out non-humans. Baxter objects, noting that the inarticulate, in the form of children and the disabled, have interests that must be protected by the community of justice without ever having competing conceptions of the good.[49] The possession of a competing conception of the good is not therefore a necessary requirement.

Introducing the inarticulate into the community of justice, however, does not solve the question of representation. Baxter asserts that a solution could lie in allowing the inarticulate, whether human or non-human, to have their interests protected through the use of proxies.[50] Members of the community of justice may act as guardians, speaking on behalf of non-human nature and future generations.[51] Including non-humans within the community does not impede Barry's notion of neutrality. This is because the act is independent of notions of the 'good' and is concerned instead with criteria of admission into the community.[52] This position, Baxter emphasizes, does not require an equality of consideration between human and non-human interests;

---

44   Barry, B. (1995), *Justice as Impartiality* (Oxford, Clarendon Press).

45   Baxter, B.H. (2000), 'Ecological Justice and Justice as Impartiality', *Environmental Politics* 9, 43–64, 45; See also Baxter, B.H. (2004), *A Theory of Ecological Justice* (London, Routledge).

46   Baxter 2000 (n. 45 above), 49.

47   Ibid. 50.

48   Ibid. 52.

49   Ibid. 53–54.

50   Ibid. 55.

51   Ibid.

52   Ibid. 57.

instead it merely requires that the interests of the non-human world be considered prior to decisions being made.[53]

Robyn Eckersley, noting the failure of Rawls's theory of democratic liberalism to respond to green concerns, takes her starting point from Habermas.[54] The latter aims for justice as part of an open discourse through which principles of justice are constructed. This dialogical approach contrasts with the monological approach employed by Rawls where the fundamental principles that govern the basic structure of society are developed in advance of the discourse. Dialogical theories aim to reach mutual agreement around the superior argument, insisting that all those affected by decisions should be active participants in the dialogue leading up to those decisions.[55] Central to discursive democracy is free and open communication[56] which also means, however, that there is no guarantee for fundamental principles to be recognized.

Eckersley identifies ecological interests as 'generalizable' interests; these are by definition not sectional, private or selfish. Generalizable interests, she notes, are protected by the mode of communication presupposed by the discourse ethic where political communication is not distorted by power imbalances.[57] In such an environment generalizable interests may be successfully defended publicly against narrower sectional interests. Although more likely to deliver green policies, Eckersley notes that this is not guaranteed and is in fact contingent upon the superiority of ecological arguments.[58] In light of this, her key concern is to bring non-human and future generations into the speech community. This requires tackling the hurdle of the discourse ethic's resistance to allowing individuals to speak on behalf of others. In the context of nature, Eckersley responds to the criticism that humans are incapable of speaking for non-humans because of the risk that humans are unable to move beyond their inherent anthropocentrism.[59] This obstacle is overcome by an acknowledgement that the conventional view of nature is incomplete, culturally filtered and provisional and therefore care must be taken in relations with nature.[60] Thus provided humans recognize their limits, the criticism is not deemed fatal. Once she achieves a place for non-humans in the speech community, she turns to the more difficult issue of how their speech can be represented.[61] Eckersley advances a form of trusteeship held by humans for nature.[62] She also introduces another option, a basic rule of thumb to guide human decision-makers away from damaging the environment that cannot speak for itself. She suggests the precautionary principle as such a norm, that would achieve

---

53   Ibid. 60.
54   Eckersley, R. (2004), *The Green State: Rethinking Democracy and Sovereignty* (Cambridge, MA, MIT Press), 141.
55   Ibid. 116.
56   Ibid.
57   Ibid. 117.
58   Ibid. 119.
59   Ibid. 123.
60   Ibid. 125.
61   Ibid. 127–138.
62   Ibid. 125.

the goal of protecting those incapable of participating actively in the discourse.[63] The principle she argues could be entrenched constitutionally to ensure systematic consideration is granted to ecological concerns.[64]

Konrad Ott[65] is concerned with 'solving the demarcation problem' between discourse ethics (Habermas 1991) and non-anthropocentric morality. Rejecting Angelika Krebs's argument that discourse ethics inevitably reaffirms anthropocentrism,[66] Ott makes a distinction between the core of discourse ethics and the set of applications which constitute a concept of morality. The core of discourse ethics has to be justified by reflexive arguments as a prerequisite for free and open communication. In the field of applications, however, different patterns of argumentation can be used. They can express a rationality in whatever form and can be ethical in whatever conceptual framework. Thus, the core of discourse ethics is compatible with non-anthropocentric reasoning. Habermas himself has increasingly employed wider concepts of rationality for his concept of discourse ethics, including morality, aesthetics, intuition, empathy and even reverence for life.[67]

David Schlosberg seeks an expansion of the meaning of justice to encompass notions of recognition and participation.[68] This wider conception of justice flows from an acceptance of Iris Young's view of injustice as institutionalized domination and oppression.[69] Injustice can only be solved through a tripartite notion of justice to include recognition, participation and distribution rather than through questions of distribution alone. Recognition is central to human dignity and involves acknowledgement that some communities are unfairly affected by environmental degradation.[70] Principally, this includes non-human communities. Political participation through inclusive decision-making can therefore achieve more just outcomes for all communities in question.[71] The three elements are depicted as inextricably linked. A lack of recognition leads to a decline in participation precluding the possibility of equitable distribution. Distributional equity cannot

---

63   Ibid. 135.

64   Ibid.

65   Ott, K. (2008), 'Solving the Demarcation Problem', in Westra, L., Bosselmann, K. and Westra, R. (eds), *Reconciling Human Existence and Ecological Integrity* (London, Earthscan), 39.

66   Krebs, A. (2000), 'Das teleologische Argument in der Naturethik', in Ott, K. and Gorke, M. (eds), *Spektrum der Umweltethik* (Marburg, Metropolis), 67–80.

67   Ott 2008 (n. 65 above), 48; Habermas, J. (1991), *Erläuterungen zur Diskursethik* (Frankfurt, Suhrkamp); Habermas, J. (2001), *Die Zukunft der menschlichen Natur* (Frankfurt, Suhrkamp).

68   Schlosberg, D. (2003), 'The Justice of Environmental Justice: Reconciling Equity, Recognition, and Participation in a Political Movement', in Light, A. and De-Shalit, A. (eds), *Moral and Political Reasoning in Environmental Practice* (Cambridge, MA, MIT Press), 79; See also Schlosberg, D. (2007), *Defining Environmental Justice* (Oxford, Oxford University Press).

69   Schlosberg 2003 (n. 68 above), 81.

70   Ibid. 81.

71   Ibid. 92.

occur in the absence of recognition and participation.[72] Schlosberg is hopeful that a tripartite view of justice might help to forge links between environmental and social justice movements.[73]

Schlosberg aims for an extended version of social justice to encapsulate elements of ecological justice. He sees a lack of recognition of the natural world as a central cause of humankind's oppression and domination of nature.[74] The application of recognition to nature is defended against the objection that the self-worth of nature cannot be adversely affected by non-recognition because of an inability to suffer psychological harm.[75] He rejects the objection on the grounds that the central element of injustice on the grounds of non-recognition is not the psychological harm suffered by the subject but rather the lack of recognition by a society.[76]

Schlosberg then identifies two directions that might be taken in constructing a recognition of nature. The first would respond to nature's intrinsic worth. The second would focus on the importance of nature for humans now and in the future.[77] Recognition of nature can, according to Schlosberg, be defended under both heads. Physical abuse, he notes, is a key element of disrespect that in turn is an aspect of non-recognition. The physical abuse of nature's integrity through human action is a sign of a lack of recognition.[78] The further ground of human self-interest in the face of environmental catastrophe also supports respect for nature.[79] Schlosberg further considers whether recognition of nature is a violation of the central principles of liberalism. He contends that including the natural world in the community of justice through recognition does not devolve into a notion of the good.[80] It is the procedural conditions for justice that are affected, rather than any particular definition of the good. Certain notions of the good may no longer be available, for example, destruction of species. However, Schlosberg stresses the difference between limiting notions of the good to protect justice for all and requiring adherence to a specific conception of the good.[81]

After defending the expansion of recognition to nature, Schlosberg turns to considering how participation of nature might be achieved. For this to occur he asserts it is necessary to consider the institutional biases against nature that lead to

---

72   Ibid. 96.

73   Schlosberg, D. (2005), 'Environmental and Ecological Justice: Theory and Practice in the US', in Eckersley, R. and Barry, J. (eds), *The State and the Global Ecological Crisis* (Cambridge, MA, MIT Press) 97–116.

74   Schlosberg, D., 'Three Dimensions of Ecological Justice', Paper prepared for Political Research Annual Joint Sessions, Grenoble, France, 6–11 April 2001, available at <http://www.essex.ac.uk/ecpr/events/jointsessions/paperarchive/grenoble/ws6/schlosberg.pdf> (draft), 12, accessed 1 December 2007.

75   Ibid. 14.

76   Ibid. 14.

77   Ibid.

78   Ibid. 15.

79   Ibid.

80   Ibid. 16.

81   Ibid. 16.

unequal distributions.[82] These can be overcome through more open and participatory structures of environmental decision-making that encourage greater political engagement and a diversity of viewpoints to be expressed.[83] An increase in diverse human participation is thus central. Nature has a role to play more directly through signals that may be interpreted by humans. Schlosberg follows Dryzek's call for a consideration of nature's signals expressed through disruptions to ecological integrity. Examples include global warming, droughts, floods, species' extinction and other physical occurrences.[84] Schlosberg thus places emphasis on involving both the human and the non-human world in the justice discourse.

Validity for the Legal Discourse

The various approaches to eco-justice all aim to integrate the non-human world in environmental decision-making. Principally, the integration can be pursued either through the ethical discourse or through the justice discourse. It would not matter if both discourses would equally lead to better decision-making. But do they? According to Rawls, justice is based on a commonly agreed discourse and, therefore, facilitated by institutions (law and governance). Ethics, on the other hand, reflect comprehensive ideals that cannot *per se* be communicated through institutions. Ethics may inform justice, but cannot guide decision-making in ways that the institutions of justice are offering.

The categorical distinction between justice and morality determines all liberal conceptions of justice. The inherent assumption is that issues of justice are different from, and superior to, moral values. Justice represents the values of an assumed public rationality (individual freedom and rights) that disassociates itself from a morality of compassion and empathy. No matter how important the non-human world may be regarded as by the members of the *justitia communis*, it stays outside the *justitia communis*.

Rawls has always been clear about this exclusion: '[the] status of the natural world and our proper relation to it is not a constitutional essential or a basic question of justice'.[85] While he acknowledged 'duties' in this regard, he described them as 'duties of compassion and humanity' rather than duties of justice. Any 'considered beliefs' to morally include the non-human world 'are outside the scope of the theory of justice'.[86] Rawls's 'original position' cannot assume a morality without a moral agent. An agent can be a person living today or a fictitious person living tomorrow, but has to be a 'person'. Interestingly, non-human entities can qualify as persons – a corporation for example – but non-human living entities cannot.

Liberal approaches to justice are bound to exclude the non-human world. Attempts to extend justice can be made, of course, but they can hardly be marginalized as a

---

82   Ibid.
83   Ibid. 17.
84   Ibid.
85   Rawls, J. (1993a), *Political Liberalism* (New York, Oxford University Press), 246.
86   Rawls, J. (1993b), *A Theory of Justice*, rev. edn (New York, Oxford University Press), 448.

'green twist' (Wissenburg 1999) of Rawls's original position. Why should Rawls or any liberal throw the baby out with the bath water? Their anthropocentric bias prevents them from expanding inter-human justice to include 'inter-species' justice.[87]

Legal theories of justice traditionally suffer from avoiding the moral debate. This explains why no legal theory has ever embraced an eco-ethical concept of justice.[88]

Ethics, in whatever form, ought to be understood as informing ANY idea of justice. There is no justice without some underpinning morality, just as there are no human rights without ethical assumptions. For example, whether or not property rights should be defined to include obligations is not a matter of the 'law', but of the ethical reasoning underpinning it. Most constitutions define private property as a combination of guaranteed individual freedom and limiting social responsibility. Property cannot be protected in abstract, but only in a social context.

It can, therefore, be reasoned that human rights are limited not solely by their social context, but also by their ecological context. Individual freedom is determined not just by laws of society, but also by laws of nature. The ecological approach to human rights has influenced human rights theory, constitutional development in Germany[89] and international law.[90]

It is worth comparing the justice debate to the rights debate. Fundamentally, the objections against a concept of ecological justice are the same as those against ecological rights.

Central to the liberal idea of justice has been the liberal conception of rights. And even though neither Rawls nor Ronald Dworkin nor any other leading liberal theorist of justice has argued for the possibility of extending rights to animals and plants,[91] it would be possible to do so, at least structurally. Whether or not non-human entities like animals or plants can have rights is a matter not so much for lawyers, but for philosophers to decide. From a legal perspective rights can be attributed to all sorts of entities like, for example, companies and states. There is no legal reason to confine rights to the sphere of human beings. From a philosophical perspective, on the other hand, the issue becomes more complex and many have written about nature's rights. There is, in fact, a tradition of rights for nature since ancient times (Stoic School) including such names as Spinoza, Leipnitz, Goethe, Schopenhauer, Bentham and, in our days, Jonas, Meyer-Abich or, with respect to animals, Singer. All of these are philosophers, but it was Christopher Stone[92] who triggered a broad public debate on

87 [See below, 99].

88 There are remarkably few lawyers among the theorists of ecological justice.

89 Bosselmann, K. (2001), 'Human Rights and the Environment: Redefining Fundamental Principles?', in Gleeson, B. and Low, N. (eds), *Governance for the Environment* (London, Palgrave), 118–34.

90 Taylor, P. (1999), 'From Environmental to Ecological Human Rights: A New Dynamic in International Law?', *Georgetown International Environmental Law Review* 10, 309–97; Bosselmann, K. (1998), *Ökologische Grundrechte* (Baden-Baden, Nomos).

91 With the possible exception of Joseph Raz who writes: 'Rights ground requirements for action in the interest of other beings', in Raz, J. (1986), *The Morality of Freedom* (Oxford, Oxford University Press), 180.

92 Stone, C.D. (1972), 'Should Trees Have Standing?', *South California Law Review* 45, 450–501; See also Stone, C.D. (1987), *Earth and Other Ethics: The Case for Moral*

rights and social change. After more than 30 years of debate it can be concluded that nature's rights are compatible with liberal theory, but less so with ecologism. The reason is that liberal rights perpetuate a core of individual freedom that is hostile to ecological responsibilities. These are perceived as potentially threatening individual freedom, thus requiring a 'balance' between rights and obligations. Only if the core itself is understood to include an ecological dimension, can the balancing act become an act of realizing human potential. In other words, the nature's rights discourse is meaningless unless reflective of the ideological context of its participants.[93] Stone himself always thought of his advocacy for nature's rights as a call for profound ethical and social change, not for updating anthropocentrism.[94]

If a mere extension of rights is not delivering ecological decision-making, why should an extension of justice help us? Applying liberal principles more widely only reinforces individualism and anthropocentrism. Instead, we need to examine the principles themselves. If they are 'blind' to the ecological context of human existence, they need to be reconsidered.[95]

Relating Ethics to Law and Justice

Relating the ethical discourse to the legal discourse automatically raises a few important questions. One is: can ecocentric reasoning penetrate legal theory moulded in anthropocentrism?

A second question (discussed further below) concerns the ethical discourse itself. How could ecocentrism be accepted as a basis for eco-justice? Can it be 'assumed' in a similar way as anthropocentrism and individual autonomy are assumed for the liberal approach to eco-justice? Or can ecocentrism only be achieved through a public dialogue, i.e. promoted as discourse ethic?

Anthropocentrism or Ecocentrism?

As indicated above, the relationship between anthropocentrism and ecocentrism is not one of gradual difference, but one of paradigmatic dichotomy.[96] This does

---

Pluralism (New York, Harper and Row).

93 Bosselmann,   K.   (1995a),   'Nichtanthropozentrische   Erweiterung   des Umweltverwaltungsrechts?', in Nida-Rümelin, J. and Pfordten, D. (eds), *Ökologische Ethik and Rechsttheorie* (Baden-Baden, Nomos), 201, 203.

94 Stone 1972 (n. 92 above).

95 It should be conceded though that we need more debate on whether a liberal approach to eco-justice is a step in the right direction or further way from it. See Almond, B. (1995), 'Rights and Justice in the Environmental Debate', in Cooper, D. and Palmer, J. (eds), *Just Environments – Intergenerational, International and Inter-Species Issues* (London, Routledge), 1–17, 6; Bosselmann, K. 1999 (n. 2 above), 39.

96 The dichotomy is, of course, not total. There are many shades of anthropocentric ethical positions as well as a variety of non-anthropocentric, i.e. ecocentric positions. For a comprehensive discussion of the various positions and their importance for environmental law see Bosselmann, K. (1995b), *When Two Worlds Collide: Ecology and Society* (Auckland, RSVP), 317–40; Bosselmann, K. (2002a), 'The Concept of Sustainable Development'

not mean, however, that they cannot co-exist in law. As we will see later, there are many examples in domestic and international environmental law where traditional anthropocentric instruments of 'natural resource' management co-exist with modern ecocentric instruments of ecosystem management. Environmental law can accommodate ethical pluralism in much the same way as it can reflect political pluralism.

There are limits however. In law, 'moral pluralism'[97] can work only as long as there are no irreconcilable positions at the same 'conflict-level'.[98] An example is the protection of endangered species and biodiversity. The Giant Panda can well be protected for anthropocentric reasons, while biodiversity requires an ecocentric approach, at least, potentially. Ecosystem protection, including humans and the non-human world, is a lot more complex, ultimately requiring us to think ecocentrically. As J. Baird Callicott observed, you cannot follow a 'happy-go-lucky moral pluralism' switching from Bentham's utilitarianism to Schweitzer's reference for life, on to Leopold's land ethic and back to Kant's categorical imperative without reflecting on what might be at stake.[99]

What then is at stake when considering 'ecological' rather than 'environmental' justice?

The complexities of the 'environment' are best captured by the term ecological integrity. It reflects the view that there are natural processes necessary to maintain the Earth's life support systems that humans and all life depend on. In other words, it is not the environment, but the interactions between the various life forms – including human beings – we should be concerned with. This view is not 'objective', but reflective of observations made. Those observations include ecological systems, evolutionary and other processes. But no observation can be made without some form of reflection or interpretation. We could ask, for example, why there is scientific interest in the environment at all? Quite obviously, environmental research followed the realization that there are problems with the environment (pollution, degradation, etc.). There is a close nexus between perceptions, observations and reflections. Our knowledge will never be 'objective', i.e. purely based on 'facts'.[100]

We can conclude, therefore, that notions such as 'ecological integrity' or 'ecological sustainability' express perceptions of ecological realities, not ecological reality itself. But so does the notion 'environment'! It expresses the perception of something surrounding us: we are here, the environment is the 'other'. The adjective 'environmental' does not connect both spheres in a way the adjective 'ecological' does. This term acknowledges not only the complexities of the natural world, but also the fact that humans are part of it.

---

in Bosselmann, K. and Grinlinton, D. (eds), *Environmental Law for a Sustainable Society* (Auckland, New Zealand Centre for Environmental Law), vol. 1, 89–96.

97   Stone 1987 (n. 92 above).

98   Ibid. 205.

99   Callicott, J.B. (1990), 'The Case Against Moral Pluralism', *Environmental Ethics* 12, 99–115, 112.

100   Mackey, B. (2004), 'The Earth Charter and Ecological Integrity', *Worldviews* 8, 76–92; Bosselmann 1995b (n. 96 above), 291–315.

Such terminological differences are not only of linguistic interest, they are also reflective of an ethical dilemma. The Western anthropocentric tradition, in which the term 'environment' was moulded, is reinforced if we continue to perceive human–nature complexities as merely 'environmental' relationships. Closer to the – scientific! – truth would be to speak of ecological relationships. The former favour ethical anthropocentrism, the latter ethical ecocentrism.

The conceptual differences between environmental and ecological terminologies may well disappear over time. For many, 'environmental' relationships are already perceived as holistic, ecological relationships. It is important though to avoid confusion. As long as there is no generally accepted view that the 'environment' includes the communality of humans and the natural world, it is better to distinguish between environmental (anthropocentric, liberal) and ecological (ecocentric) approaches to justice.

The ecological approach to justice is based on ecocentrism. But how can this basis be communicated? Why should it have more validity than the anthropocentric basis?

Ontological or Discourse Ethic?

All theories of justice claim validity for modern democratic societies. In fact, the justice discourse has, for a long time, assumed that justice and democracy are mutually reinforcing. A democratic society is more 'just' than an undemocratic society and will strive for justice as a key objective. Likewise, justice can only flourish in an open democratic society and will encourage ever more democratic forms of decision-making.

The assumption has been that the members of society are equipped with individual freedoms and equal rights. Each member has the same right of access to justice and decision-making. The members of society alone are entitled to organize their political system. Consequently, the system of justice results from a – hypothetical – dialogue between free individuals.

This liberal approach to justice found its best expression in Rawls's theory. Its central constructs of 'original position' and 'veil of ignorance' allow for a justice concept without presupposed ethics. The appropriate rules of justice will emerge from this hypothetical dialogue. As no one is to know their individual future fate, each member will aim for rules that are acceptable to everyone. The individual fear of ending up as one of the worst-off members of society ultimately shapes the rules and values of justice.

The geniality of Rawls's theory lies in guaranteeing basic equal rights (the 'first principle of justice') without attempting to further define what these rights may entail. Such a non-ethical, 'secular' approach reinforces the pluralism of modern liberal, democratic society. The only 'problem' is that only liberal, democratic society benefits from Rawls's theory of justice. Effectively, Rawls's theory locks society into a permanent liberal future; any move towards a more communitarian or collectively organized society ends at the barrier of individual rights.

The 'hidden' liberal ideology in Rawls is widely criticized. The question is, however, how far this critique needs to be taken to accommodate the concern for the environment. Is it sufficient to add some basic concerns such as a 'no harm principle' or 'savings principle' to an otherwise untouched veil of ignorance? Or is it necessary to unveil the assumed ignorance?

Structurally, these questions can be answered by examining the means under which socially binding norms are identified. They can either be identified through reasoning (monological or ontological approach) or through public debate (dialogical or discourse approach). Monological approaches to justice presuppose some idea of how society ought to be organized. Presupposed basic individual rights and democratic principles are examples of monological approaches. These rights and principles can also be part of dialogical approaches; however, the assumption here is that they primarily result from a public dialogue (Habermas 1991) rather than from their inherent persuasiveness. The difference that discourse ethics is trying to make is the rejection of liberal theory as the only valid basis of society.

Politically, the discourse principle aims for emancipation from the liberal state. In doing so it emphasizes the role of procedure and rationality. In Habermas's words: 'According to the discourse principle, just those norms deserve to be valid that could meet with the approval of those potentially affected, in so far as the latter participate in rational discourses'.[101] As shown above, Habermas considers environmental ethics as a legitimate part of a rational discourse. So can the rational discourse, at least, potentially include the non-human world?

This is not likely according to Rawls and his environmentally minded followers.[102] They all stand in the tradition of Modernity that accepts anthropocentric rationality as the sole guide for ethics. It can be found in utilitarianism, Kant's categorical imperative and contemporary contractual theory. In this tradition, the ethical discourse will always follow the rational discourse, for example, of justice.

The argument against ethical rationalism is that it has not stopped us from fundamentally changing the conditions that make human life possible. The development of modern civilization and technology has altered human behaviour in such a dramatic way that traditional ethics can no longer provide necessary orientation. In Einstein's famous words: 'The means for solving a problem cannot be the same as the ones that brought about the problem in the first place'.

Nobody has formulated the ecological challenge towards ethics and justice more precisely than Hans Jonas. In his seminal text on the imperative of responsibility,[103] Jonas showed that the key assumptions made by traditional ethics are no longer valid, namely that the *conditio humana* is unchangeable, that human behaviour can be calculated on the basis of rational analysis and that human responsibility is limited to social context (ethics and justice). These assumptions expressed themselves in a rigid anthropocentrism that entirely excluded the non-human world from ethical and moral categories.[104]

---

101   Habermas, J. (1996), *Between Facts and Norms: Contributions to a Discourse Theory of Law and Democracy*, tr. Rehg, W. (Cambridge, Polity Press), 127.

102   [See above, 81–85].

103   Jonas, H. (1984), *The Imperative of Responsibility: In Search of an Ethics for the Technological Age*, tr. Jonas, H. and Herr, D. (Chicago, University of Chicago Press).

104   Ibid. Forword.

If we assume, therefore, that the current ecological crisis has its roots in the *conditio humana* of European civilization[105] we need to revisit both the procedural and normative aspects of ethics and justice. Neither monological nor dialogical approaches are in themselves sufficient; the former tend to be too presumptuous, the latter too unambitious.

Following the concepts of self-reflective discourses on justice proposed by Young,[106] Tully[107] and Kingwell[108] we need a justice discourse that allows for both openness and transformation. Considering the systemic character of the crisis that we are in we should be able to incorporate ecocentric values when thinking about justice without assuming that they can be inscribed like individual rights have been inscribed in modern theories of justice. I call this the ecological justice discourse.

Robyn Eckersley's concept of 'green justice'[109] adequately describes this discourse. Her attempt to ecologically transform traditional ideas of justice not against, but within the context of democracy and the state is convincing. Justice is an inherent part of the modern democratic state and yet in urgent need of revision. Liberal justice theories ignore the close nexus between liberalism, anthropocentrism and environmental degradation. They cannot capture the ecocentric concept of justice. Equally, we cannot rely on mere discourse ethic. No matter how 'free of tyranny' this discourse is, it will not itself produce the required ecological discourse.

The main limitations of discourse ethics are its own anthropocentric roots. If this tradition behind the rational discourse goes unquestioned we will not be able to move beyond the Kantian and contractualist framework. In this framework, only moral agents count. From a non-anthropocentric perspective the limitation of moral theory to *agents* is unacceptable; rather it is enough that a being receives recognition as a moral *subject*. That such beings may not have the capacity to communicate (directly) or reciprocate moral favours should not matter.

What should matter, instead, is the capacity of humans to not only act for themselves, but also for those who cannot act for themselves. In this way potentially all beings affected by environmental decisions, but not actively participating in the moral dialogue, would have their voice heard. Whether this concern for 'the others' requires institutional representation – for example, through agencies of

105   Bahro, R. (1994), *Avoiding Social and Ecological Disaster: The Politics of World Transformation: An Inquiry into the Foundations of Spiritual and Ecological Politics*, tr. Clarke, D. (rev. edn, Bath, Gateway), Part II, Ch. 6; Bosselmann 1995b (n. 96 above), 71.

106   Young, I. (1990), *Justice and the Politics of Difference* (New Jersey, Princeton University Press).

107   Tully, J. (1995), *Strange Multiplicity: Constitutionalism in an Age of Diversity* (Cambridge, Cambridge University Press).

108   Kingwell, M. (1995), *A Civil Tongue: Justice, Dialogue and the Politics of Pluralism* (University Park, PA, Pennsylvania State University Press).

109   Eckersley, R. (2001), 'Green Justice, the State and Democracy', available at <www. arbld.unimelb.edu.au/envjust/papers/allpapers/eckersley/home.htm>, accessed 1 December 2007; see also Eckersley 2004 (n. 54. above).

guardianship[110] or trusteeship[111] – or whether it can be internalized in some other way[112] is an interesting subject of debate. For the acceptance of ecocentrism and eco-justice, however, it is only necessary to internalize what has remained externalized in theories of justice to date.

Defining Ecological Justice

Having seen how environmental ethics can influence and shape the justice discourse, we can now focus on contents. What does the concept of ecological justice entail? Can it be sufficiently defined to provide guidance for the development of environmental law?

The proximity of ecocentrism to ecological sustainability is the most promising way for a workable theory of ecological justice. As shown in Chapter 2, environmental law is increasingly influenced by the concept of sustainable development. And despite the ongoing debate on how this concept could be defined, there is a recognizable consensus on some of its core ideas. These ideas form part of the concept of ecological justice. To become a truly ecological concept, justice needs to reach out into the non-human world. As we will see, the 'missing link' in both the sustainable development debate and the justice debate is the recognition of ecological integrity. It is not enough to care for humans living today and those living tomorrow when the natural processes that sustain life are at risk. There is a need to identify and recognize the ethical and legal importance of ecological integrity.

The famous Brundtland definition[113] contains two ethical elements that are widely accepted as being essential to the idea of sustainable development:

1. concern for the poor (intragenerational justice or equity); and
2. concern for the future (intergenerational justice or equity).

The concern for the poor includes all social and minority groups discriminated against by the social and economic system. Conflicts between rich and poor, white and non-white, Western lifestyle and indigenous cultures and North and South belong here. The environmental justice movement in the United States has its origins in those conflicts. The concern for the poor (social justice) represents the social

---

110   On guardianship models used or proposed in several countries see Bosselmann, K. (2002b), 'A Legal Framework for Sustainable Development' in Bosselmann, K. and Grinlinton, D. (eds), *Environmental Law for a Sustainable Society* (Auckland, New Zealand Centre for Environmental Law), vol. 1, 145–61, 151–2.

111   On trusteeship models in international law see Taylor, P. (1998), *An Ecological Approach to International Law* (London, Routledge), 283–5; Sand, P. (2002), 'Trusteeship for Common Pool Resources: Zur Renaissance des Treuhandbegriffs im Umweltvölkerrecht', in Schorlemer, S. (ed.), *Praxishandbuch UNO* (Berlin, Springer), 201–33.

112   On the importance of the precautionary principle in this regard see Eckersley 2004 (n. 54 above), 136–7; and Bosselmann 2002b (n. 110 above), 153; see also further below.

113   Sustainable development is 'development that meets the needs of the present without compromising the ability of future generations to meet their own needs'.

dimension of ecological justice and can usefully be termed *intragenerational justice*. The importance for a theory of ecological justice is to determine the relationship between intragenerational and intergenerational issues as there are competing claims and priorities.

The concern for future generations is a familiar feature in international and environmental law. Since Edith Brown Weiss's influential classic text,[114] rights and interests of future generations have been incorporated in international agreements and national legislation. The notions of 'justice for future generations' and *intergenerational justice* may be less familiar, but both have become increasingly popular in recent years, obviously reflecting a broader acceptance of the idea that justice spans past, present and the future. One of the issues to be discussed here is whether intergenerational justice is to be seen as an anthropocentric concept or whether it has ecocentric implications.[115]

However, these two elements leave us with a crucial question. What do they mean with respect to the environment? If we are to share environmental goods and burdens fairly among those living today and also leave something for the future, what exactly do we preserve? The integrity of the planetary ecosystem (the 'natural stock') or our current knowledge about natural resources and ways to use them (the 'capital stock')?

For reasons of principle we are unable to determine the needs of future generations. Only more or less informed guesses are possible about the options that future generations may justifiably expect. The reasonable choice, therefore, is for a duty to pass on the integrity of the planetary ecosystem as we have inherited it (ecological integrity). Uncertainty requires precaution, and there seems no better precautionary measure than assuming that future generations would like the planetary ecosystem to be as bountiful as we have found it.

And yet, such an obvious duty is neither suggested by the Brundtland definition nor favoured by governments or big business. As their morality is confined to traditional social ethics, they do not consider it important for an environmental ethic to incorporate nature or the planetary ecosystem. Instead, the standard view is that society, economy and environment are somehow of equal importance. As a result, sustainable development is perceived as a balancing act between economic, social and environmental goals with trade-offs as a necessary outcome. There is no guidance that could ensure, for example, a preference for ecological sustainability or the needs of future generations. Without such guidance, policies may become more integrated, but they will make little difference to existing unsustainable patterns of production and consumption.

---

114 Brown Weiss, E. (1989), *In Fairness to Future Generations* (New York, Transnational).

115 Tladi, D. (2007), *Sustainable Development in International Law* (Pretoria, Pretoria University Press), 45; Tladi, D. (2002), 'Of Course for Humans: A Contextual Defence of Intergenerational Equity', *South African Journal of Environmental Law and Policy* 9, 177–92; Bosselmann, K. (2007), 'Strong and Weak Sustainable Development: Making the Difference in the Design of Law', *South African Journal of Environmental Law and Policy* 13, 14–23.

This business-as-usual approach is commonly known as 'weak sustainability'. If associated with moral obligations to the future, weak sustainability policies consider it our sovereign decision what kind of assets, 'stock' or legacy we wish to leave for future generations. It could be the 'natural stock', but it also could be its substitute, for example, the knowledge to alter the natural stock (e.g. through genetic engineering). Such a 'capital stock' gift to the future implies the very possibility of destroying the planet's conditions of life. We simply don't know and possibly never will.

To preserve the integrity of the planetary ecosystem is the only reliable alternative and may, in fact, be a desirable goal for most people. However, this goal needs moral and legal recognition in order to become 'internalized'. Only then will the 'weak' become the 'strong'; and the missing link in the Brundtland definition will be found. A third element, therefore, needs to be added to the two mentioned above:[116]

concern for the non-human natural world (interspecies justice or equality).

The concern for the non-human natural world is in the centre of environmental ethics. The question here is which idea of justice could accommodate such concerns best. As we have seen, from a liberal, anthropocentric point of view the non-human world is outside the *justitia communis*. From an ecocentric point of view *justia communis* includes both the human and non-human world. It may already be seen as an important step that today there is vivid discussion of 'justice for the non-human world'[117] and 'interspecies justice'.[118] This term has affinities with intra- and intergenerational justice. The inclusion of an elaborate concept of interspecies issues is certainly crucial for a theory on ecological justice as distinct from mere social justice.

Interspecies justice (or concern for the non-human natural world) is not mentioned in the standard definitions of sustainable development or in any of the international agreements related to sustainable development (e.g. 1992 Rio Declaration, Agenda 21, 2002 Johannesburg Declaration). However, this may be due to the fact that the anthropocentric notion of sustainable development has dominated the political debate. In literature, environmental or ecological sustainability has emerged as an ecocentric conception that includes the recognition of intrinsic values. This ecocentric conception stands against a technocentric conception of sustainability[119] promoted by the UNCED process and most governments. As Andrew Dobson has

---

116 Bosselmann 2002b (n. 112 above), 145, 147–56.

117 Almond 1995 (n. 95 above), 18.

118 Ibid. 15; see also Palmer, J. (1995), 'Just Ecological Principles?', in Cooper, D. and Palmer, J. (eds), *Just Environments – Intergenerational, International and Inter-Species Issues* (London, Routledge), 31–2; Johnson, A. (1995), 'Barriers to Fair Treatment of Nonhuman Life', in Cooper, D. and Palmer, J. (eds), *Just Environments – Intergenerational, International and Inter-Species Issues* (London, Routledge),165; and Hayward, T. (1997), 'Interspecies Solidarity: Care Operated Upon by Justice', in Hayword T. and O'Neill, T. (eds), *Justice, Property and Environment* (Aldershot, Ashgate).

119 See Pearce, D. (1993), *Blueprint 3: Measuring Sustainable Development* (London, Earthscan), 18–19.

observed, the ecocentric approach marks the point at which economists leave the sustainability debate.[120]

At the heart of the conflict between the technocentric and the ecocentric approaches lies the ambivalence of the word sustainability[121] which – not unlike the words love or justice – causes uniform agreement, yet great confusion as to what it actually means. It is widely accepted though that ultimately there are just two possible meanings. One starts from the argument that 'sustainability is not enough'.[122] The concern here is that sustainability too readily denotes stasis and equilibrium whereas life is about change and growth. From this perspective development is the important dynamic element making sustainable development a concept which ensures lasting economic growth and energy production or – in the words of the World Bank – simply a 'development that lasts'.[123] Opposed to this philosophy of more-is-better lies the philosophy of different-not-more. It asks the question 'sustainability of what?' The concern here is that sustainable development simply means to sustain the Western way of life at the expense of the poor and of future generations. To be sustained are not the economic, but the ecological conditions, requiring substantial changes in economy and society. From this perspective the central issue is environmental sustainability[124] and not development, making a concept of a sustainable society preferable perhaps to sustainable development.[125]

The (anthropocentric) market-based approach and the (ecocentric) equity approach to sustainable development[126] represent different political agendas. It is, therefore, not possible to *a priori* define the relationship between sustainable development and ecological justice. However, an analogy can be drawn between both concepts. 'Ecological' can be understood as modifying 'justice' in much the same way as 'sustainable' can be understood as modifying development. On this basis, the only permissive paths to development are those that are ecologically

---

120   Dobson, A. (1996), 'Environmental Sustainability: An Analysis and Typology', *Environmental Politics* 5, 401–28, 416.

121   Mitcham, C. (1996), 'The Sustainability Question', in Gottlieb, R. (ed.), *The Ecological Community* (New York and London, Routledge), 359–79; also Beckerman, W. (1994), 'Sustainable Development: Is it a Useful Concept?', *Environmental Values* 3, 191–209; and Pezzey, J. (1992), 'Sustainability: An Interdisciplinary Guide', *Environmental Values* 1, 321–62.

122   Ruttan, V. (1988), 'Sustainability is Not Enough', *AJAA* 3 (2–3), 128–30.

123   World Bank, *World Development Report 1992: Development and the Environment* (New York, Oxford University Press), 34.

124   United Nations Conference on Environment and Development (Rio de Janiero, 3–14 June 1992; 31 ILM 814), introductory note by Brown Weiss, E.; see also Draft International Covenant on Environment and Development (March 1995; World Conservation Union (IUCN), Environmental Policy and Law Paper No. 31, Rev. 2), Article 1: 'The objective of this Covenant is to achieve environmental conservation and sustainable development by establishing integrated rights and obligations'.

125   See Brown, L. (1981), *Building a Sustainable Society* (New York, WW Norton) and the annual *State of the World* reports of the Worldwatch Institute on 'progress toward a sustainable society', <http://www.worldwatch.org/taxonomy/term/38>, accessed 1 December 2007.

126   Dobson 1996 (n. 120 above), 423.

sustainable. Likewise, the only permissive paths to justice are those that recognize ecological sustainability.

The importance of ecological sustainability cannot be overestimated. For example, it gives further guidance to the meaning of intragenerational justice. Concern for the poor, essentially the support for development in the Third World, is a high priority. In fact, since the 2000 UN Millennium Goals and the 2002 Johannesburg Summit, there is an agreed consensus among states that eradication of poverty and economic development in poor countries should have priority over other goals associated with sustainable development. Such priority cannot, however, mean allowing unsustainable development. Ecological sustainability requires discouraging of unsustainable energy sources, for example, as opposed to sustainable, i.e. renewable energy sources. The ecological integrity of planetary systems should be a benchmark for any form of development.

In the same way interspecies justice should be a benchmark for any form of ecological justice. As it implies the recognition of the intrinsic value of the non-human natural world, interspecies justice can be very effective in law.

The example of the law related to biotechnology can illustrate this.

At the international level, biotechnology became the subject of international law through the 1992 Convention on Biological Diversity.[127] Along with a general trend in recent international environmental law, the Biodiversity Convention takes the approach of ecosystem protection (i.e. protecting entire habitats rather than individual species as such).[128] It does so by introducing (in its Preamble) an 'intrinsic value of biological diversity', in addition to 'the ecological, genetic, social, economic, scientific, educational, cultural, recreational and aesthetic values of biological diversity and its components'. This is recognition of the distinction between (ecocentric) intrinsic values and (anthropocentric) instrumental values of the environment. Article 19 of the Biodiversity Convention calls for the contracting states to take legislative measures towards controlling biotechnological research activities. The problem is that the Convention, like most treaties, leaves the means of implementation totally to the discretion of states.

At the municipal level, several countries have introduced such controlling legislation, among them Germany with its Gene Technology Law (*Gentechnikgesetz*) of 1990. Such legislation regulates details of notification and licensing of genetically modified products (such as, for example, the release of those products into the environment), but it always does so on the basis that there is a fundamental right to conduct genetic engineering in the first place. The principle of free production and sale is the rule, any restrictions are the exception. The burden of proof, therefore, is not with the producer introducing a new risk potential (e.g. for biological diversity), but with the general public represented, for example, by expert commissions such as the Environmental Risk Management Authority in New Zealand or various

---

127   Convention of Biological Diversity 1992 (Rio de Janiero, 5 June 1992; 31 ILM 818 (1992))

128   Kiss and Shelton 2004 (n. 5 above), 11.

ethical commissions in the UK. Whether or not activities of genetic engineering are acceptable is determined by weighing up social costs and benefits. The problem is that such social costs and benefits are exclusively determined by values of human utility. As there are no intrinsic values of ecosystems and their components to be considered, the only relevant counterweight comes from possible infringements of rights of health protection, i.e. human health risks associated with genetically modified products. Once these concerns are met nothing could stop genetic engineering from fundamentally altering the natural genetic structures. The recognition of nature's intrinsic values would have, at least, the potential to rectify such grave imbalances and provide the counterweight necessary for more balanced decisions.

Apart from those reasons for improving the *results* of decision-making, there are reasons for improving the *basis* for decison-making. Before today's complete breakdown of ethical agreement – the 'moral catastrophe' (MacIntyre 1981) modernity has witnessed[129] – there was once a well-established ethical basis for decision-making. We could go back to the beginning of modern civilization and remind ourselves of the close links between ethics, justice and law. In Aristotle's quest for justice,[130] for example, we find the directive of 'the good life' and some of the virtues associated with such a life: imagination, openness, empathy. The life of someone who is unimaginative, closed-up and rigid *may* be 'good', of course, but it is reasonable to suggest that such a person is indifferent towards fellow humans and non-human life around them. Being alienated and cut off from the natural world has become a dominant feature of modernity. So, resemblence to Aristotle's 'good life' and its virtues may help to rediscover, for example, what empathy is all about.[131] It seems to me that there is no contradiction between empathy for the natural world and the good life of humans. There is a lot to suggest that the tragedy of modernity has been to create a dichotomy between autonomy and dependence; we badly need a more sophisticated sense of human interdependence.[132] A sense of interconnectedness may well be the key for future decision-making.

Ultimately, there are no 'reasons' why we should adopt intrinsic values of nature. Ethical reasoning can only help by clarifying some arguments. Ultimately, we have to rely on our ability to learn from past experience and see things differently. Interspecies justice is a new concept which cannot simply be attached to the idea of justice. The 'barriers to fair treatment of non-human life'[133] are deeply embedded in the psyche of modernity and can only be removed by revisiting both the idea of justice and our place in the world.

---

129   MacIntyre, A. (1981), *After Virtue; A Study in Moral Theory* (London, Duckworth).

130   Urmson, J. (1988), *Aristotle's Ethics* (Oxford, Oxford University Press); Hardie, W. (1980), *Aristotle's Ethical Theory* (Oxford, Oxford University Press).

131   A connection between ecocentric awareness and 'the good life' is made by Cooper, D. (1995), 'Other Species and Moral Reason', in Cooper, D. and Palmer, J. (eds), *Just Environments – Intergenerational, International and Inter-Species Issues* (London, Routledge), 137, 145–7.

132   Tronto, J. (1993), *Moral Boundaries: A Political Argument for the Ethic of Care* (London and New York, Routledge), 101.

133   Johnson 1995 (n. 118 above), 165–79.

The elements of ecological justice are now before us, and we may have enough building blocks for a possible theory. However, since the building blocks sit on ecocentric foundations they cannot claim to be universally acceptable. They may be useful though to facilitate further discussion. The aim should, in any case, be to work on a conception which does justice to humans and nature alike.

The idea of eco-justice is, however, far from being speculative. As the last chapter will show, its elements are already present in law. Recent developments in international and municipal law give an indication of how eco-justice is being implemented in environmental policy and legislation.

Ecological Justice in Environmental Law

There is no example of a conscious, deliberate attempt to implement ecological justice in legislation. This should not be surprising, however, considering not only the novelty of this idea, but also the function of justice as a fundamental concern. Law will always reflect some sense of justice, but not in a direct manner as it would, for example, implement a specific principle. What is 'fair' and 'just' depends on the subject matter of the law, its ideological and political context and the importance that may be given to specific aspects of justice. The practical use of eco-justice is its focus on the wider ecological context that legislation, administration and judicial review operate in. If we are conscious of the three principles of eco-justice, we have a more informed view of how environmental law should be designed and interpreted.

Some examples in the area of domestic and international environmental law can illustrate how eco-justice has already influenced legislative developments.

New Zealand's Environmental Legislation

As mentioned earlier, New Zealand's environmental legislation is often hailed as one of the world's most advanced[134] although this may be truer for its ambition rather than its current operation. The Environment Act 1986 and the Conservation Act 1987 espouse an ecocentric approach by providing holistic definitions for the 'environment' (including humans and nature) and recognition of 'intrinsic values of ecosystems'. The actual implementation of this approach occurred during a reform period between 1984 and 1991 when the Resource Management Act (RMA) was adopted with the 'ethic of sustainable management'[135] at its core. The RMA is a

---

134 See Chapter 2. One political historian described the reforms in New Zealand as 'the greatest changes anywhere, anytime, in any democracy', quoted in May, P. et al. (eds) (1996), *Environmental Management and Governance: Intergovernmental Approaches to Hazards and Sustainability* (London and New York, Routledge), 43; see also Kloepfer, M. and Mast, E. (1995), *Das Umweltrecht des Auslandes* (Berlin, Duncker & Humblot), 35, 301–10; and Robinson, N., (1997), *IUCN (The World Conservation Union) Newsletter*, July–September, 3.

135 Upton, S. (Minister for the Environment) (1994), 'The Resource Management Act, Section 5: Sustainable Management of Natural and Physical Resources', Keynote speech to the Second Annual Conference of the Resource Management Law Association, Wellington, New Zealand, October 1994; see also Grundy, K. (1995), 'Sustainable Management: A

remarkable example of efforts to incorporate sustainable development into law as it integrates socio-economic and environmental issues.[136] It is also a good example of applied ecological justice as it provides for a definition of sustainable management reflective of the three principles of eco-justice.

To begin with, the RMA identifies two major functions of 'sustainable management', the *management* function and the *ecological* function. The object of the management function is the use, development and protection of resources which include social, economic and cultural well-being and the health and safety of people and communities. The object of the ecological function is sustaining the potential of resources, safeguarding the life-supporting capacity and avoiding adverse environmental effects as described in paragraphs (a), (b) and (c) of section 5(2) RMA.

The word management is a neutral term and sets no particular values and priorities. However, use, development and protection of resources are equally important and to be managed simultaneously. It is in this context that both intragenerational and intergenerational justice appear. Under section 5(2), decisions regarding the distribution of resources must sustain the needs of future generations. Distribution does not follow a quantification of such needs, but rather the criterion of absolute equality.[137] Despite the reference to 'the reasonably foreseeable needs', resource distribution is determined by sustaining the potential of resources *in view of* future needs, rather than by those needs themselves. This would be consistent with the ethical approach to intergenerational equity as defined above,[138] thus *potentially* consistent with an ecocentric interpretation of future generations.[139] The proper reading of future generations – whether reduced to humans or life as a whole – depends very much on the reading of the overall ethics shaping sustainable management.

While the management functions are essentially anthropocentric in character, the ecological functions are not. Paragraphs (b) and (c) – and arguably even (a) (concerning future generations) – of section 5(2) RMA express long-term considerations on the basis of ecocentrism. However, crucial for this interpretation is the proper meaning of the little word 'while' between management and ecological functions[140] and the

---

Sustainable Ethic?', Paper for the Third Annual Conference of the Resource Management Law Association, Wellington, New Zealand.

136   The overall objective of Agenda 21 (and elaborated in eight chapters) is to restructure decision-making so that consideration of socio-economic and environmental issues is fully integrated. Apart from this, the Brundtland definition of sustainable development (n. 113 above) was considered by the legislators, but the term 'management' was preferred to 'development' because of a lesser degree of social equity and global distribution issues inherent in the RMA.

137   [See above, 64].

138   [See above, 35].

139   Bosselmann, K. (1999), 'Justice and the Environment: Building Blocks for a Theory of Justice', in Bosselmann K. and Richardson, B. (eds) *Environmental Justice and Market Mechanisms: Key Challenges for Environmental Law and Policy* (London, Kluwer Law International), 30, 46–8.

140   The relevant passage of section 5(2) reads: 'Managing the use ... of ... resources in a way ... which enables people and communities to provide for their ... well-being ... *while*

proper linking of the remaining sections (6 to 8). There has been a lot of debate as to whether 'while' introduces the ecological functions as being superior or subordinate to the management functions.[141] The former idea is referred to as 'ecological bottom-lines', the latter as an 'overall judgment approach'.

A grammatically correct interpretation would suggest 'while' to mean 'by way of' rather than a non-committed 'and' or 'also'. Consequently, all management functions are to be conducted in an ecologically sound way as defined in paragraphs (a) to (c).[142] Interspecies equity can be seen as being addressed in these paragraphs and further defined in the 'other matters' of section 7 for which decision-makers are to 'have particular regard'. Among theses matters are the 'intrinsic values of ecosystems' (paragraph d) and the 'concept of "katiakitanga"' (paragraph a) which is defined as the 'exercise of guardianship' and, in relation to a resource, includes the 'ethic of stewardship based on the nature of the resource itself'. Recognition of intrinsic values and reference to guardianship clearly espouse an ecocentric understanding of interspecies equity.

Leaving aside the number of difficulties that have occurred in practice,[143] the RMA is by no means value-free. The ethics and values expressed in the definition of sustainable management carry the elements of eco-justice.[144] Intragenerational, intergenerational and interspecies equity are all being addressed, albeit not always in a language clear enough to define the interrelations between them and ethical principles underlying them. Yet ecocentrism clearly defines the ecological functions, thereby helping us to understand that environmental justice is, essentially, justice for those who cannot speak for themselves.

Ecosystem Regimes and Management

In a broader sense, environmental law has been increasingly informed by ecology and ecosystem approaches. The science of ecology is concerned with a number of

---

(a) sustaining the potential of ... resources (for) ... future generations; *and* (b) safeguarding the life-supporting capacity ... *and* (c) avoiding, remedying, or mitigating any adverse effects' (emphasis added).

141 See Pardy, B. (1997), 'Planning for Serfdom: Resource Management and the Rule of Law', *New Zealand Law Journal*, 69–72; Williams, D.A.R. (1997), *Environmental and Resource Management Law* (Wellington, Butterworths), 75.

142 Bosselmann and Taylor 2005 (n. 139 above), 117. It should be noted that various policy papers seem to favour more of a 'balance' between management and ecological functions; on the other hand, Environment Court Judge Kenderdine speaks of 'cumulative safeguards' supporting the view expressed here: *Foxley Engineering Ltd v. Wellington City Council* (W 12/94).

143 These led, in November 1998, to proposals of the government to remove some of the procedural obstacles within the RMA.

144 See also Gunn, A. and McCallig, C. (1997), 'Environmental and Social Justice in an Urban Setting – Sustainable Management and the New Zealand Resource Management Act 1991', Paper for the Conference 'Environmental Justice', University of Melbourne, 1–3 October.

levels of relationships between organisms and their environment.[145] These levels include genes, individuals, populations, communities, ecosystems and landscapes to, ultimately, the entire Earth. Historically, ecologists confined their work to one ecological level and had very few forms of interaction with other levels. However, there has been a growing trend of ecologists to look at more complex interactions among a number of levels. This trend is increasingly reflected in environmental laws becoming more integrated and wider in their approach.

Generally speaking, environmental laws and regulations seek to control human interactions with ecosystems. If these interactions are seen from the viewpoint of an autonomy of human life, then both the scientific explanation through ecology and the control mechanisms through law will be comparatively simple: the environment appears separate from humans and will be protected insofar as is necessary to maintain the quality of human life. Essentially, the autonomous self is retained intact and only affected by certain impacts (pollution, water scarcity, etc.).

If, by contrast, the interactions are seen from the viewpoint of relationships between humans and nature, then the environment appears as more fundamental to human life. The more the complexities of these relationships are understood, the more emphasis will be given to interdependences between humans and nature. This second viewpoint of humans regards the self as a momentary concretization of components and processes of a large ecosystem. In other words, the boundaries between the human and the natural spheres disappear.

Currently the development of environmental law is somewhere between the first and the second viewpoint. Recent trends towards ecosystem regimes suggest that environmental law may reach this second viewpoint at some point in the future. The ecosystem approach to legal regimes[146] is visible in the use of terms such as 'carrying capacity', 'biodiversity' or 'ecosystems', but also in the increased use of normative principles such as integrated environmental management, the precautionary principle or the concept of (ecologically) sustainable development.

The rise of ecosystem regimes[147] has been gradual, but steady, and during the 1990s reached a certain stage of maturity. The New Zealand RMA 1991 is one prominent example, but most OECD countries concerned themselves with similar integrated and ecosystemic environmental law regimes. Typically, ecosystems such as marine and coastal areas, regional areas (e.g. water catchments, forestry), biodiversity, even climate systems are covered by more comprehensive statutes and regulations involving broader concepts ('sustainability'), integrated management plans and new

---

145 For good overall discussions of ecology and its various subdivisions see McIntosh, R. (1985), *The Background of Ecology: Concept and Theory* (Cambridge, Cambridge University Press); Begon, M., Harper, J. and Townsend, C. (1996), *Ecology: Individuals, Populations and Communities*, 3rd edn (London, Blackwell Science); and Bush, M. (2000), *Ecology of a Changing Planet*, 2nd edn (Upper Saddle River, NJ, Prentice Hall).

146 See, for example Young, O. (1982), *Resource Regimes: Natural Resources and Social Institutions* (Berkeley, University of California Press); Ostrom, E. (1990), *Governing the Global Commons: The Evolution of Institutions for Collective Action* (Cambridge, Cambridge University Press).

147 Brooks, R. et al. (2002), *Law and Ecology: The Rise of the Ecosystem Regime* (Aldershot, Ashgate), 2.

institutions. Several countries including the Netherlands, Scandinavia, Germany and Australia adopted new general environmental codes informed by sustainable development. In addition, most European countries created institutional frameworks for sustainability in the form of Green Plans (Netherlands, Sweden, France) and National Strategies (UK, Germany, etc.). Similar strategies were adopted in Canada, the USA, Australia and New Zealand.

Ecosystem regimes have an inherent tendency towards greater comprehensiveness and higher complexity. Along with this tendency the interdependences between humans and nature will be more obvious and with them the need for ecological justice. Exactly what principles the various ecosystem regimes follow is not always clear. There is no consistent set or hierarchy of guiding principles. Sometimes the emphasis is on management issues, sometimes on ecological sustainability; one law may emphasize resource use, another precaution and a third, the needs of future generations. It is obvious, however, that all the various principles touch upon eco-justice issues. In some way or other they are all concerned with issues of intragenerational, intergenerational and interspecies justice. The further development of ecosystem laws – and environmental law in general – would greatly benefit from a new jurisprudence with eco-justice at its core.[148]

International Environmental Law

Ecosystemic approaches to law-making are also known in international law. Treaty law recognizes intrinsic values of the natural world, for example, with respect to biological diversity.[149] To some extent the Convention on Biological Diversity,[150] has pushed the frontiers of international environmental law further than ever before, expressly incorporating the precautionary principle,[151] common concern of humankind[152] and common but differentiated responsibilities,[153] and including reference to intrinsic values.[154]

Apart from the Biodiversity Convention, many environmental treaties have recognized the intrinsic value of the biosphere. Examples include the 1991 Protocol on Environmental Protection amending the 1959 Antarctic Treaty,[155] the 1982

---

148 Such new, ecological jurisprudence suggests that legal rules should be closely tied to scientific findings ('ecology') and ethical norms ('justice'); Fondacaro, M. (2000), 'Toward an Ecological Jurisprudence Rooted in Concepts of Justice and Empirical Research', *UMKC Law Review* 69, 179–96.

149 Bosselmann 1999 (n. 2 above), 50.

150 Convention on Biological Diversity 1992 (n. 127 above).

151 Note that reference to the precautionary principle is in the Preamble only, an express article included in the fifth negotiating draft was removed. See Boyle, A. (1996), 'The Rio Convention on Biological Diversity', in Bowman, M. and Redgwell, C. (eds), *International Law and the Conservation of Biological Diversity* (London, Kluwer Law International), 33–49, 37.

152 Convention on Biological Diversity 1992 (n. 127 above), Preamble.

153 Ibid., Article 6; but see generally, Boyle 1996 (n. 151 above), 44–5.

154 Convention on Biological Diversity 1992 (n. 127 above), Preamble.

155 Recognizing the intrinsic value of the Antarctic ecosystem.

World Charter for Nature,[156] and a number of treaties related to the preservation
of ecosystems[157] and endangered species.[158] Mention should also be made of the
32 so-called Alternative Treaties negotiated by several hundred non-governmental
organizations at the 1992 Earth Summit in Rio de Janeiro. For example, in
response to the failed UN Earth Charter – substituted by the anthropocentric Rio
Declaration – the NGO delegates adopted a first version of the Earth Charter based
on an ecocentrically defined responsibility for the earth.[159] The 1995–2001 IUCN
Draft International Covenant on Environment and Development follows a similar
approach; e.g. Article 2 establishes the principle of respect for all forms of life.

Since the mid 1990s the Earth Charter had been negotiated by many hundreds of
civil society groups representing the world's cultural, ethnic and religious diversity.
In 2000 it was adopted in The Hague and has since been endorsed by a number
of international organizations and states. The Earth Charter aims for more than
global environmental precaution, it provides a comprehensive ethical framework
for sustainable development. The Charter takes a systemic view on peace, security,
social and ecological justice, human rights and democracy. None of these can be
achieved without the other, and they all need to be developed to reflect the Charter's
idea of sustainable development. Its principles are, therefore, guidelines for the
entire way in which nations and people ought to conduct their affairs. This makes
the Earth Charter a suitable constitution for a new world order.

The Charter provides the 'values and principles for a sustainable future'.[160] It
assumes the validity of the Brundtland definition for sustainable development, but
clearly identifies the 'missing link' of ecological sustainability mentioned above.
Principle 1 ('Respect Earth and life in all its diversity') reflects the core of ecological
justice: 'Recognize that all beings are interdependent and every form of life has value
regardless of its worth to human beings'. The first set of Principles, 1 to 4 on 'Respect
and care for the community of life' and Principles 5 to 8 (on 'Ecological integrity')
further describe interspecies justice, which has been missing in the general discourse
on sustainable development. Principles 9 to 12 ('Social and economic justice') and

---

156   Stating, in its preamble, that 'Every form of life is unique, warranting respect
regardless of its worth to man and, to accord other organisms such recognition, man must be
guided by a moral code of action'.

157   For example, the Ramsar Convention on Wetlands 1971 and the Berne Convention
on the Conservation of European Wildlife and Natural Habitats 1979 recognize the intrinsic
value of wild fauna and flora.

158   Examples include the Convention on International Trade in Endangered Species of
Wild Fauna and Flora 1973 and the recently amended regime on whaling; D'Amatao, A. and
Chopra, S.K. (1991), 'Whales: Their Emerging Right to Life', *AJIL* 85, 21–62.

159   Published in Pacific Institute of Resource Management (ed.) (1992), 'Commitment
for the Future: The Earth Charter and Treaties agreed to by the International NGOs and
Social Movements', Paper presented to the International NGO and Social Movements Forum
Conference, Wellington, New Zealand, 11 June.

160   Earth Charter Commission (2002), 'Earth Charter: Values and Principles for
a Sustainable Future, The Earth Charter Initiative', <www.earthcharter.org>, accessed 1
December 2007.

13 to 16 ('Democracy, non-violence, and peace') then describe intragenerational and intergenerational justice.

It is clear that international environmental law is increasingly recognizing aspects of interspecies justice. In this respect, pure anthropocentrism is already being compromised along ecological lines.[161] It is timely, therefore, to develop sustainability law along the lines of ecological justice.

---

161 Redgwell, C. (1996), 'Life, The Universe and Everything: A Critique of Anthropocentric Rights', in Boyle, A. and Anderson, M.R. (eds), *Human Rights Approaches to Environmental Protection* (New York, Clarendon Press), 87.

Chapter 4

Ecological Human Rights

Introduction

Human rights thinking has responded to sustainability concerns for quite some time. The most prominent development has been the increased recognition of a distinct human right to a clean and healthy environment. Other developments include the increased importance of the right to life and physical well-being in cases of local pollution or, more recently, widespread climate change litigation[1] where individual rights and collective rights (e.g. of indigenous peoples) play a central role.

The interdependence between human rights and environmental protection is increasingly recognized in international and domestic law. However, fundamentally each area remains to be guided by its own legal regime. Human rights law is concerned with the protection of individual well-being; environmental law is concerned with the protection of collective well-being. There is, at present, little penetration between both regimes although this may change over time.

From an ecological perspective, the separation of human rights law and environmental law is not in itself a problem. What does matter, however, is the rationality underpinning each. What form or paradigm of rationality applies when we think of human rights (or the environment, respectively)?[2] Paradigms of rationality have associated value systems. Value systems refer to the relative importance assigned to competing values. If, for example, human welfare is perceived to be superior to environmental welfare, conflicts will be resolved in a manner that favours human needs (in their entire spectrum) over environmental needs. As a consequence, the degree to which the assumed superiority is used will determine the degree of environmental protection. And if this superiority manifests itself in unrestrained property rights, economic growth and unfettered utilitarianism, then clearly the environment will suffer.

An economic rationality of human rights favours individual and material values over collective and immaterial values. An ecological rationality of human rights, on the other hand, would not necessarily reverse this order, but question its underpinning utilitarianism. The economic rationality assumes the Greco-Christian position that everything on Earth is for the sole use of humankind.[3] Claims for an intrinsic value

---

1 See, for example, Smith, J. and Shearman, D. (2006), *Climate Change Litigation* (Adelaide, Presidian Legal Publications); Taylor, P., *Climate Change Litigation: A Catalyst for Corporate Responses* (forthcoming).

2 See Hancock, J. (2003), *Environmental Human Rights* (Aldershot, Ashgate), 15–33.

3 Ibid. 22.

of nature tend to be dismissed as irrational and non-quantifiable.[4] To the present time, human rights have not posed a challenge to economic rationality. Their individual entitlements are compatible with individualism and materialism. In a similar vain, the design of environmental law has not been inconsistent with economic rationality. Ultimately, the relationship between human rights and the environment is determined by its prevailing rationality, not by legal reasoning *per se*.

This chapter reviews the development of environmental human rights[5] in terms of their legal recognition and their underpinning philosophies. As we will see, two distinct developments have emerged, one following the traditional rationality of protecting individual freedoms, the other following the new rationality of protecting the environment. While both developments have been influencing each other, their underpinning rationalities have only in part been complementary.

To a degree, the concern for the protection of human rights and the concern for protection of the environment are mutually reinforcing. Human rights and environmental law are both needed to provide better human living conditions. To some other degree, however, the protection of individual rights has been counterproductive to the protection of the environment. Property rights, in particular, have not been conducive to achieving ecological sustainability. The overarching importance of sustainability requires a more coherent approach, in essence, a comprehensive, unifying regime of human rights and obligations. How such a regime would look and what real progress has already been made is the main concern of this chapter.

Reviewing the international development of environmental human rights we need to first consider two methodological aspects. One aspect concerns the different levels of international law, national law and supranational (European Union) law. Each of these levels follows its own human rights approach, and at the national level we find a whole variety of legal traditions expressing rather different concepts of human rights.[6] But there is also a degree of commonality, in particular, with relation to environmental issues. The global nature of environmental issues makes for a certain similarity of human rights responses. This allows a meaningful stock-taking of contemporary environmental rights.

The other aspect concerns the various forms in which human rights are applied to the environment. They can be used to indirectly combat environmental degradation (threats to existing human rights), they can be used for more effective environmental decision-making processes (procedural environmental rights), and they can be used to more directly enforce environmental protection (human right to a healthy environment). While each of these approaches emphasizes rights and entitlements, a fourth approach emphasizes human responsibilities. Here we ask how duties

---

4   Gowdy, J. (1999), *Coevolutionary Economics: The Economy, Society and the Environment* (Boston, Kluwer); Bosselmann, K. (1995), *When Two Worlds Collide: Society and Ecology* (Auckland, RSVP).

5   Understood here as a generic term for human rights with an environmental dimension whether by content or by context.

6   See generally Steiner, H.J., Alston, P. and Goodman, R. (eds) (2007), *International Human Rights in Context* (Oxford, Clarendon Press); Birnie, P. and Boyle, A. (2002), *International Law and the Environment*, 2nd edn (Oxford, Oxford University Press); Donnelly, J. (2007), *International Human Rights*, 3rd edn (Boulder, CO, Westview Press).

towards the environment can best be formulated to protect and preserve ecological sustainability.

From a sustainability perspective, rights need to be complemented by obligations. Merely advocating environmental rights would not alter the anthropocentric concept of human rights. If, for example, property rights continue to be perceived in isolation and separation from ecological limitations, they will reinforce anthropocentrism and encourage exploitative behaviour. We need to consider, therefore, a human rights theory based on non-anthropocentric ethics. Ecological approaches to human rights are, in fact, not just theoretical. They have informed constitutional debates and international documents as we will see later.

The overall aim of this chapter is to show the dramatic influence that the principle of sustainability is having on human rights and our thinking about individual freedom, property and the interrelations between rights and responsibilities.

International Recognition of Environmental Human Rights

As indicated, the international regime for the protection of human rights has developed differently from the protection of the environment. The former emerged from the post World War II recognition of fundamental freedoms, in particular the 1948 Universal Declaration of Human Rights. The latter emerged from the recognition of a global environmental crisis, in particular the 1972 Stockholm Conference on the Human Environment. Over the last 30 years, both regimes have increasingly influenced each other; however, it is important to recall the basic function of human rights in the context of international law.

With the adoption of the Universal Declaration of Human Rights in 1948 states have, for the first time in their history, restricted their own sovereign powers. With the further adoption of the International Covenants of Economic, Social and Cultural Rights, and Civil and Political Rights, in 1966, states have recognized restrictions to their internal governance.[7] Effectively, this has curtailed state sovereignty. No state can exempt itself from the fundamental obligation to protect an individual's life and dignity. This cannot merely be perceived as a voluntary self-restriction of states, but rather as the consequence of the nature of human rights. They are rooted in natural law reflecting universal principles of morality.[8]

In so far as human rights reflect a rule of basic necessity, the same could be said about the environment. Theorists of environmental law have often referred to environmental protection as a matter of basic necessity[9] resembling natural law thinking.[10] Kiss and Shelton, for example, explain the theoretical foundations

---

7   Steiner, Alston and Goodman 2007 (n. 6 above), Chapter 3; Sloane, R.D. (2001), 'Outrelativizing Relativism: A Liberal Defense of the Universality of International Human Rights', *Vanderbilt Journal of Transnational Law* 34, 527–95, 532.

8   Sloane 2001 (n. 7 above), 542–3.

9   Coyle, S. and Morrow, K. (2004), *Philosophical Foundations of Environmental Law* (Oxford, Hart Publishing); Brooks, R., Jones, R. and Virginia, R. (2002), *Law and Ecology* (Aldershot, Ashgate).

10   See generally Bosselmann 1995 (n. 4 above), 231–7.

of international environmental law as a conglomerate of religious, philosophical and scientific as well as economic and social considerations requiring a truly interdisciplinary and integrated approach. [11] While environmental law could not be derived from an objective 'law of nature', its very existence reflects a commonly held view that the environment is indispensable. In this sense, protecting human life and dignity and protecting the environment follow the same basic concern for life.

The ethical and legal digestion of this basic concern is, of course, far from being complete. Legally, humans count for a lot more than the environment as a subject of protection. There is as yet no commonly shared view that human welfare depends on the welfare of the entire living world. It is not surprising, therefore, that the development of environmental human rights since the 1980s has been dominated by traditional anthropocentrism.

Notwithstanding this dominance, it is significant that both international human rights law and international environmental law have their origins not in treaty law, but in international conferences and soft law documents. With their respective subject matters rooted more in fundamental concerns of humanity than in negotiated interests of states, they share a certain degree of partisanship that does not sit too comfortably with state sovereignty. Most commentators see human rights law and environmental law as challenging the orthodoxy of international law rather than blending in with it. Thus, the prominence of non-binding, or soft law approaches is typical for both and also characteristic of the new ground that they have paved together, i.e. environmental human rights.

Environmental Harm and Human Rights

Whenever environmental harm occurs the enjoyment of human rights is potentially at risk. A standard situation is, therefore, the exposure of individuals to air pollution, contaminated water or chemical pollutants. The human rights approach here is 'unreservedly anthropocentric',[12] but may affect a wide spectrum of recognized human rights. The basic argument is that the environment must not deteriorate to the point where the right to life, the right to health and well-being, the right to family and private life, the right to property and other human rights are seriously impaired. In the words of ICJ Judge Weeramantry:

> The protection of the environment is ... a vital part of contemporary human rights doctrine, for it is a sine qua non for numerous human rights such as the right to health and the right to life itself. It is scarcely necessary to elaborate on this, as damage to the environment can impair and undermine all the human rights spoken of in the Universal Declaration and other human rights instruments.[13]

---

11   Kiss, A. and Shelton, D (2004), *International Environmental Law*, 3rd edn (New York, Transnational Publishers), 27.

12   Ibid. 143.

13   See case about the Gabçikovo-Nagymaros Project (37 ILM 162 (1998)) at 206, cited in Kiss and Shelton 2004 (n. 11 above).

From this observation, it seems rather obvious to consider sound environmental conditions as part of the right to life. In so far as this right protects from serious risks to human life, the source of such risks should not matter. To this end, there is an obvious link between environmental health and human health, and international human rights law had little difficulty deriving environmental rights from existing human rights such as life, well-being, private life, or property.[14] What is important though are the dynamics underlying this realization. Once the links between environmental health and human health are noticed, it becomes a problem to distinguish the two. Ecological perspectives tend to emphasize the connections between environmental degradation and human rights violations. Human rights perspectives, on the other hand, tend to maintain the differences between them, not because they would be less sympathetic to environmental courses, but for legal reasons. Why is this so?

As we will see in the following discussion of human rights cases, there is often no reasonable threshold between damages to the environment that are considered a mere nuisance in human rights terms, those damages overstepping the threshold to human rights violation and other damages causing large-scale threats to human dignity and life, yet no human rights violations.

It seems evident that a disaster like a methane explosion in a nearby municipal waste dump violates rights to life, privacy or property.[15] But less evident are cases where the impact is not so immediate and individualized, but long-term and large-scale, affecting entire populations. The main example is global warming. Typically, climate change is perceived as threatening environmental health, human health and property, but only to a lesser degree as threatening human dignity and rights. Climate change litigation is happening in many countries, and it is reasonable to expect a lot more litigation in the years to come.[16] However, the difficulties in the way of success are often insurmountable.

Apart from establishing causation, there are problems with finding the right scale and the right form of action. Seeking compensation for harm caused by global warming may be possible, for example, for the loss of homes and livestock from floods. But tort cases are rare, more often litigation is used to prevent or reduce further global warming. Such cases are directed against corporations, public authorities, governments or states but for the most part their success lies in their symbolic value; they focus on public attention and may be successful in so far as they influence governmental or corporate policies. However, the individualistic core of human rights is not conducive to this kind of action. Even where environmental and other public interest groups claim human rights violations of entire populations, the legal yardstick is the individual right to life or property. This harbours a reductionist, almost preposterous logic: the more people are threatened, the less likely are violations of human rights. This signals a dramatic gap between morality and legality of climate change, and the issue to be discussed is how to best close the gap.

---

14 Birnie and Boyle 2002 (n. 6 above), 252.
15 *Oneryildiz v. Turkey* [2004] ECHR 657 (30 November 2004).
16 See, for example, Smith and Sherman 2006 (n. 1 above).

One possibility is to insist on the superiority of morality. The argument is that human rights fundamentally reflect human dignity and life as the highest values of modern civilization. It would, therefore, not be acceptable to reject human rights protection on the grounds of their individual nature. Massive threats to human dignity and life will have to be considered as violations of individual human rights.[17]

The argument against this viewpoint is that it underestimates the power of legal rights. They are not a direct reflection, but an ideologically filtered reflection of morality. Historically and systematically, human rights have emerged from political liberalism favouring the protection of individual freedom over the protection of groups or entire populations. This 'reduced concept of freedom'[18] means that human rights only protect individualized legal positions in relative isolation from social and ecological conditions. This isolation was never complete and many environmental rights cases show that collective goods can be protected through human rights. But there are limits. If individual freedom is perceived as being threatened by collective interests – no matter how urgent they may be – it will prevail. An example in case is individual property rights dominating over social and ecological responsibilities. To close the gap between morality and legality of rights, human rights themselves need to be redefined.[19]

To test the validity of these two arguments, we will take a closer look at the development of environmental rights.[20] My argument is that environmental human rights have a role to play, in fact, some of their qualities are crucial for improving environmental protection, but they are also limited in their focus and can be counterproductive to ecological sustainability. To this end, there is a need for ecologically defined human rights.

As mentioned above, there are three different categories of environmental rights, namely threats to existing human rights, procedural environmental rights and a right to a healthy environment.

Procedural Environmental Rights

Of these, procedural rights are the least problematic. Being essentially democratic and participatory, this kind of rights aims for transparency, accountability and participation in decision-making. In so far as they allow for public engagement in environmental decision-making, they appear to strengthen concerns for ecological sustainability. However, certain limitations need to be pointed out.

---

17   Brown, D. (2008), 'The Case for Understanding Inadequate National Responses to Climate Change and Human Rights Violations', in Westra, L., Bosselmann, K. and Westra, R. (eds), *Reconciling Human Existence and Ecological Integrity* (London, Earthscan).

18   Bosselmann 1995 (n. 4 above), 226.

19   Taylor, P. (2008), 'Ecological Integrity and Human Rights', in Westra, L., Bosselmann, K. and Westra, R. (eds), *Reconciling Human Existence and Ecological Integrity* (London, Earthscan), 89.

20   For a recent account see also Collins, L. (2007), 'Are We There Yet? The Right to Environment in International and European Law', *McGill International Journal of Sustainable Development Law & Policy* 3, 119–53.

There is a long history of demanding procedural environmental rights. Principle 23 of the 1980 World Charter for Nature, for example, states that 'All persons, in accordance with their national legislation, shall have the opportunity to participate, individually or with others, in the formulation of decisions of direct concern to their environment, and shall have access to means of redress when their environment has suffered damage or degradation'.[21] Principle 13 of the Earth Charter demands to 'Strengthen democratic institutions at all levels, and provide transparency and accountability in governance, inclusive participation in decision-making and access to justice'.[22] Agenda 21 recognized that 'one of the fundamental prerequisites for the achievement of Sustainable Development is broad public participation in decision-making'.[23] And Principle 10 of the Rio Declaration set out specific aspects of procedural environmental rights.

The most advanced international instrument is the 2001 Aarhus Convention on Access to Information, Public Participation in Decision-Making, and Access to Justice in Environmental Matters.[24] As a regional convention, initiated by the United Nations Economic Commission for Europe, it was initially confined to European states; however, by the end of 2007 it had been signed and ratified by 40 primarily European and Central Asian countries and the European Union. Although still regional in scope, the significance of the Aarhus Convention is global[25] and represents the most elaborate treaty of Principle 10 of the Rio Declaration. As its title suggests, the Convention is built around the three broad themes of access to information, public participation and access to justice. However, it also contains a number of important general features. They include the mentioning in Principle 1 of the 'right of every person of present and future generations to live in an environment adequate to his or her health and well-being' as a general objective. The question in our context is to what extent this objective can actually be achieved through the Convention.

Principle 9 ('Access to Justice') provides for access to a review procedure before a court of law to 'challenge the substantive and procedural legality' (para. 2) of environmental decisions. The EU itself has introduced legislation to activate public access to environmental information[26] and public participation in the drafting of certain plans and programmes.[27] There is also a Proposed Directive on Access to Justice in

---

21   World Charter for Nature 1982 (GA Res. 37/7, UN Doc. A/37/51).

22   Available online from <http://www.earthcharter.org> (Home Page), accessed 1 December 2007.

23   Agenda 21: Programme of Action for Sustainable Development (Agenda 21) (Rio de Janeiro, 14 June 1992; UN Doc. A/Conf. 151/26 (1992) 31 ILM 874 (1992)), para. 23.2.

24   See <http://www.unece.org/env/pp/>, accessed 1 December 2007.

25   The Convention is open to accession by non-European countries, subject to approval of the Meeting of the Parties.

26   Directive 2003/4/EEC (28 January 2003), Public Access to Environmental Information.

27   Directive 2003/35/EEC (26 May 2003), Providing Public Participation in Respect of the Drawing Up of Certain Plans and Programmes Relating to the Environment.

Environmental Matters.[28] The EU legislation has been implemented in most EU member states, but there are, however, problems with the actual extent of judicial review.

Germany, for example, has extensive legislation on access to justice allowing environmental and other public interest groups to seek judicial review. When the Working Group of the Parties to the Aarhus Convention recently reviewed the implementation process it identified a number of shortcomings and drafted a 'Long-Term Strategic Plan'.[29] In the process it became clear that environmental groups hardly ever succeeded in seeking judicial review of environmental matters. In the case of Germany, a review of the national implementation process showed that NGOs were only able to claim breaches of procedural rights and individual rights. Effectively, this excluded any environmental issues such as 'ground water, nature protection or other fields of environmental protection that are ruled in common interest and not also in private interest'.[30] This means that ecological sustainability would only be dealt with in so far as individual rights are affected.

The implicit or explicit separation of public interests and (private) individual rights underpins all current regimes of environmental governance. Typically, NGOs are not perceived as representing environmental interests, but public interests. Both overlap only in part, and as only public interests can be brought to bear when access to information and decision-making are sought, any following judicial review is limited to breaches of procedural rights (of NGOs) or individual environmental rights (health, property). Environmental health and ecological sustainability fall largely through the gaps.

If the implementation of the Aarhus Convention is indicative of procedural environmental rights in general, then expectations for effective environmental protection need to be cautioned. Procedural rights are democratic rights and important as such. However, they are only a prerequisite for better environmental decision-making and do not in themselves safeguard ecological sustainability.

Threats to Human Rights

Ever since the 1972 Stockholm Declaration, making the link between environmental degradation and enjoyment of human rights, the environmental dimension of human rights has been recognized in international law and in many national jurisdictions. While there is no consistent recognition with uniform standards, it is commonly accepted today that environmental harm can cause a violation of human rights.[31]

Various human rights tribunals have noted that failure of public authorities to protect citizens from environmental harm can raise issues of human rights protection.

---

28   03/0624 final COD 2003/0246, Commission proposal for a Directive on access to justice.

29   Available at <www.unece.org/env/documents/2007/pp/ece_mp_pp_wg_1_2007_L_12_e.pdf>, accessed 1 December 2007.

30   Christian, S. (2007), in Working Group of the Parties to the Aarhus Convention, *Compilation of Responses to the Draft Strategic Plan*, 16, available at <http://www.unece.org/env/pp/LTSP/Compilation_public_comments_2007_06_01.pdf>, accessed 1 December 2007.

31   [See Weeramantry's statement above, 114].

The UN Human Rights Committee has held, for example, that the storage of nuclear waste in the community can cause a threat to individual rights to life.[32] In another case, the Inter-American Commission on Human Rights found that the Brazilian government had violated the (indigenous) Yanomami people's rights to life, liberty and personal security by failing to prevent serious environmental damage caused by mining companies.[33] These and other cases suggest a general acceptance that environmental harm can cause human rights violations of entire populations.

However, there are limitations. In the case of Ominayak v. Canada,[34] the leader of the Lubicon Lake Band of the Cree Indian Band of Alberta argued that the Band's land had been expropriated and destroyed due to oil and gas exploration. He alleged violations by the government of Canada of the Band's right of self-determination and the right to freely dispose of its natural wealth and resources and not be deprived of its own means of subsistence contrary to Articles 1 to 3 of the 1966 International Covenant of Civil, Cultural and Political Rights. For the Cree, environmental preservation is foundational for their culture and formed an integral part of their self-determination. Canada argued that the Covenant could not be invoked because it dealt with individual rights, but the Committee decided it could examine the case under other Articles of the Covenant including Article 27. It stated that 'the rights protected by Article 27 included the rights of persons, in community with others, to engage in economic and social activities which are part of the culture of the community to which they belong'.[35] However, the Committee separated those rights from concerns underlying the right to sovereignty over natural resources. It rejected the Cree's argument that the ecology of the land was their right rather than property and development rights. In a Separate Opinion, Mr Ando disagreed on the finding of violation of Article 27 saying: 'It is not impossible that a certain culture is closely linked to a particular way of life and that industrial exploration of natural resources may affect the band's way of life'.[36] This case, along with others, illustrates the difficulty of fitting the cultural-environmental context of people into their individual human rights. There are tensions between the environment as a collective concern and human rights protection.

As indigenous peoples are the first to suffer from environmental degradation, their importance is comparable to 'canaries' as Bradford Morse[37] remarks. He also

---

32   *EHP v. Canada*, Communication No. 67/1980, in United Nations Human Rights Committee (1990), *Selected Decisions of the Human Rights Committee under the Optional Protocol* (United Nations Publications), vol. 2, 20.

33   *Yanomami Indians v. Brazil Inter-Am.* (CHR 7615, OEA/Ser.L.V/II/66 doc. 10 rev. 1 (1985)), available at <http://www.cidh.org/annualrep/84.85eng/Brazil7615.htm>, accessed 1 December 2007.

34   *Ominayak and the Lubicon Lake Band v. Canada*, UN Human Rights Committee, Communication No. 167 (1984), UN Doc. CCPR/C/38/D/167/1984 as reported in Robb, C. (ed.) (2001), *International Environmental Law Reports: Trade and Environment* (Cambridge, Cambridge University Press), vol. 3, 27.

35   Ibid. 59, para. 32.2.

36   Ibid. 60 (Separate Opinion of Mr Ando).

37   Morse, B. (2008), 'Indigenous Rights as a Mechanism to Promote Environmental Sustainability', in Westra, L., Bosselmann, K. and Westra, R. (eds), *Reconciling Human*

stresses that indigenous peoples are teachers of sustainable living, holders of unique rights including environmental protection and lead activists for social change. In his recent survey of international and national cases, Morse found evidence for increased recognition of indigenous rights. However, using the legal tools available has its limitations. They can increase environmental sensitivities, but will not in themselves lead to the much needed elevation of ecological sustainability.

The European Court of Human Rights has, in a number of cases, recognized the impact that environmental harm can have on rights protected under the European Convention of Human Rights. In Onerylidz v. Turkey,[38] the applicants cited the right to life (Article 2), the right to private and family life (Article 8) and the right to peaceful enjoyment of possessions (Article 1 of Protocol No. 1) as being violated through a methane explosion at a nearby municipal waste dump. The Court largely agreed and held that Article 2 imposes 'a positive obligation on States to take appropriate steps to safeguard the lives of those within their jurisdiction'.[39] This obligation 'entails above all a primary duty on the State to put in place a legislative and administrative framework designed to provide effective deterrence against threats to the right to life'.[40]

Potentially, such reasoning could include a rights-based enforceable obligation of the state to protect the environment. Case law in this direction, however, has been restricted to severe forms of environmental pollution with a direct impact on individuals' rights. In Taskin and Others v. Turkey,[41] for example, the Court found that 'Article 8 applies to severe environmental pollution which may affect individuals' well-being and prevent them from enjoying their homes in such a way as to affect their private and family life adversely'.[42] So far, there is no indication to go further and state a general duty to protect the environment in order to satisfy human rights protection. The fact that only immediate environmental effects on human health and welfare have been considered as human rights violations suggests a restrictive interpretation. Environmental degradation is only relevant and enforceable in so far as it causes a direct and severe violation of individuals' rights.

Unfortunately, the 'silent' environmental crisis originating in singular events and single localities, but spreading to regions, ecosystems and, ultimately, the entire planet, is not being addressed by existing human rights. With respect to human rights, the 'tragedy of the commons' lies in the fact that most forms of environmental degradation are perfectly legal. Individually, human rights such as property rights represent entitlements to the use of the environment. Collectively, the exercise of

---

Existence and Ecological Integrity, (London, Earthscan), 159.

38   *Onerylidz v. Turkey* (n. 15 above).

39   Ibid. para. 71.

40   Ibid. paras. 89–90.

41   *Taskin and Others v. Turkey*, 46117/99 [2004] ECHR 621 (10 November 2004) at para. 113, citing *López Ostra v. Spain*, judgment of 9 December 1994, Series A no. 303-C, § 51.

42   *Taskin and Others v. Turkey* (n. 41 above); see also *Powell and Rayner v. United Kingdom*, 172 Eur. Ct. HR (ser. A), (1990); *Arrondelle v. United Kingdom*, App. No. 7889/77, 5 Eur. HR Rep. 118, 119 (1982) (European Commission on Human Rights) (friendly settlement).

rights leads to systemic, large-scale environmental degradation. This phenomenon is hardly being addressed through relying on existing human rights.

The Human Right to a Healthy Environment

A logical consequence from such weakness is the postulation of a new human right. A distinct human right to a healthy environment was first formulated in Principle 1 of the 1972 Stockholm Declaration: 'Man has the fundamental right to freedom, equality and adequate conditions of life, in an environment of a quality that permits a life of dignity and well-being, and he bears a solemn responsibility to protect and improve the environment for present and future generations'. Notably, the new right was coupled with a new 'solemn' responsibility. The Stockholm nexus between the right to and the responsibility for the environment probably reflected the political climate of the time. The experience of the looming environmental crisis was fresh in the minds of the public and conference delegates, leading to a covenant-type declaration of rights and responsibilities.[43]

Since Stockholm, the human right to a healthy environment has been recognized in many soft law documents and legal instruments, as well as in national constitutions and domestic court decisions. By contrast, the responsibility to protect and improve the environment was not considered to be of relevance to this new or any existing human right. The liberal concept of human rights is not conducive to accepting legal responsibilities, leaving it to environmental law to address them. Environmental laws now exist in many countries. And although general environmental codes[44] covering the various forms of human impact on the environment are still rare, environmental laws have broadened the coverage over time and continue to do so.[45] The unresolved problem is the isolation of environmental law from its wider legal and ethical context. Law in general still circles around property[46] and no fundamental concept of environmental responsibility exists. There is, at present, no overall framework to resolve conflicting values and objectives.

In the absence of a coherent legal framework, environmental responsibilities, even where they exist in law, cannot be balanced against the rights to use the environment. The question is, therefore, whether a human right to a healthy environment would make a difference and considerably raise the level of environmental protection.

---

43   The Declaration contains responsibilities with respect to the conservation of natural resources (Principles 2 to 7), specifics about the implementation of environmental protection (Principles 8 to 25) and the foundations for the further development of international law (Principles 21 to 26). Together with the comprehensive 'Action Plan for the Human Environment' the Declaration had immense importance for the development and evolution of environmental law. See Kiss and Shelton 2004 (n. 11 above), 60–64.

44   For example, the US National Environmental Protection Act 1969 or the New Zealand Resource Management Act 1991.

45   Broadly speaking, while laws in most OECD countries cover land use, water and air quality, waste, chemicals, and aspects of wildlife relatively well, other areas such as soil fertility, biodiversity, climate, and renewable energy are much less legislated, not to mention the linkages between social and economic activities and ecological integrity.

46   Bosselmann 1995 (n. 4 above), 51–62.

In the 15 years between the Stockholm Conference and the appearance of the Brundtland Report in 1987, there was no significant progress towards a human right to a healthy environment. The Brundtland Report itself made no connection between the new idea of sustainable development and such a right. It did, however, receive a mention in the Proposed Legal Principles for Environmental Protection and Sustainable Development accompanying the Report.[47]

In contrast to Principle 1 of the Stockholm Declaration, Principle 1 of the 1992 Rio Declaration states that 'Human beings are at the centre of concerns for sustainable development. They are entitled to a healthy and productive life in harmony with nature'. Despite a UN Resolution from 1990 in favour of a human right to a healthy environment,[48] the Rio Declaration avoided unequivocal language. Principle 1 of the Rio Declaration has been accepted with no reservations at several subsequent UN conferences[49]

The most comprehensive UN report was written in 1994 by Fatma Ksentini, the Special Rapporteur on Human Rights and the Environment. The report[50] highlights the reciprocal relationship between rights and duties with respect to the environment. It also makes the case that a human right to a healthy environment, while important in itself, should not be conceived in lieu of a duty to protect the environment. As an appendix to the report, the 1994 Draft Declaration of Principles on Human Rights and the Environment reflects this duality. Principle 2 states that 'All persons have the right to a secure, healthy and ecologically sound environment' and specifically categorizes it as a human right. Principle 21 states that 'All persons, individually and in association with others, have a duty to protect and preserve the environment'.

Building upon the Ksentini Report, the Bizkaia Declaration from 1999 recognizes, in Article 1, that 'Everyone has the right, individually or in association with others, to enjoy a healthy, ecologically balanced environment ... [which] may be exercised before public bodies and private entities'. The Preamble to this Declaration refers to Principle 1 of the Stockholm Declaration, the Rio Declaration's recognition of environmental entitlement, regional treaties, UN General Assembly Resolution 45/94 and other international documents as evidence for an emerging human right.

---

47   Experts Group on Environmental Law of the World Commission on Environment and Development (1988), *Environmental Protection and Sustainable Development, Legal Principles and Recommendations* (Dordrecht, Martinus Nijhoff Publishers), 14.

48   United Nations GA Res. 45/94, UN Doc. A/RES/45/94 (1990), 'Need to Ensure a Healthy Environment for the Well-Being of Individuals'.

49   UN Conference on Population and Development 1994, the World Summit for Social Development 1995, the Second UN Conference on Human Settlements, and the OAS Hemispheric Summit on Sustainable Development 1997, as cited in Lee, J. (2000), 'The Underlying Legal Theory to Support a Well-Defined Human Right to a Healthy Environment as a Principle of Customary International Law', *Columbia Journal of Environmental Law* 25, 283–346.

50   UN ESCOR Commission on Human Rights, Sub-Commission on Prevention of Discrimination and Protection of Minorities (1994), *Review of Further Developments in Fields with which the Sub-Commission Has Been Concerned, Human Rights and the Environment: Final Report Prepared by Mrs Fatma Zohra Ksentini, Special Rapporteur*, UN Doc. E/CN.4/ Sub.2/1994/9.

The recently adopted UN Declaration on the Rights of Indigenous Peoples[51] takes a different approach to the right to environment, placing it in a wider context. The right of indigenous peoples to a vibrant, safe and sustainable environment is defined as part of cultural integrity and the right to self-determination. In her recent book on the rights of indigenous peoples, Laura Westra shows the importance of ecological integrity in this context. It can be seen as combining the elements of cultural integrity and self-determination and therefore as the basis for rights and duties towards the environment.[52]

Also in 2007, the French Head of State and senior UN representatives adopted the Paris Appeal calling for the adoption of a 'Universal Declaration of Environmental Rights and Duties'.[53] The motivation behind this Appeal is the concern for environmental ethics leading to duties to complement any rights.

At regional level, some treaties have formally recognized the right to a healthy environment. Article 24 of the African Charter on Human and Peoples' Rights states that 'All peoples shall have the right to a general satisfactory environment favorable to their development'.[54] For the Americas, the Additional Protocol to the American Convention on Human Rights in the area of Economic, Social and Cultural Rights (the Protocol of San Salvador) recognizes the right to a healthy environment in Article 11. Article 2 requires states to promote the protection, preservation, and improvement of the environment.[55]

As far as the European Union is concerned, neither the 2000 Charter of Fundamental Rights nor the (rejected) 2004 European Constitution provide for a right to a healthy environment. However, section 37 of the Charter[56] and section 97 of the Constitution[57] correspondingly acknowledge the importance of environmental protection: 'A high level of environmental protection and the improvement of the quality of the environment must be integrated into the policies of the Union and

---

51   UN Declaration on the Rights of Indigenous Peoples, A/RES/61/295, adopted 13 September 2007 available at <http://www.iwgia.org/graphics/Synkron-Library/Documents/InternationalProcesses/DraftDeclaration/07-09-13ResolutiontextDeclaration.pdf>, accessed 1 December 2007.

52   Westra, L. (2007), *Environmental Justice and the Rights of Indigenous Peoples* (London, Earthscan).

53   Yves Lador, 'Time for a Universal Declaration on Environmental Rights', <http://partnerships4planet.ch/en/environmental-rights.php>, accessed 1 December 2007.

54   African Charter on Human and Peoples' Rights 1981 (Banjul Charter).

55   Protocol of San Salvador 1988 (Article 11: 'The States Parties shall promote the protection, preservation, and improvement of the environment'). This obligation of states to adopt the measures necessary to provide for the rights listed in the Protocol is somewhat limited by the proviso in Article 1, which provides that states' available resources and degree of development are to be taken into account.

56   Charter of Fundamental Rights of the European Union, adopted 7 December 2000, 2000/C 364/01, available at <http://www.europarl.europa.eu/charter/pdf/text_en.pdf>, accessed 1 December 2007.

57   Treaty Establishing a Constitution for Europe 2004 (Doc. 2004/C310/01), available at <http://eurlex.europa.eu/LexUriServ/site/en/oj/2004/c_310/c_31020041216en00410054.pdf>, accessed 1 December 2007.

ensured in accordance with the principles of sustainable development'. By contrast, the draft constitution of 1994[58] had recognized an individual right.[59]

In 2003, the Council of Europe Parliamentary Assembly adopted a report on the environment and human rights[60] and recommended the drafting of an additional protocol to the European Convention on Human Rights in which the right to a healthy environment would be clearly defined. The report raises some interesting issues. First, it justifies the need for an internationally recognized human right to a healthy environment with 'defects in international environmental law' described in three respects: the conflict between economy and ecology, the fragmentation of environmental law and ongoing non-compliance.[61] The purpose of a new human right is, therefore, to improve law and governance rather than to provide better individual protection. Second, to overcome the defects a radical new approach is needed: 'From an economic perspective, the concept of sustainability now needs to be extended to include the new idea of "strong sustainability" – which is based on the assumption that there is a core stock of natural capital, that cannot be replaced and must therefore be kept constant over time'.[62] By relating (strong) sustainability to human rights the report offers broader considerations than are usually made with respect to environmental rights. For example, it acknowledges 'the obvious difficulties involved in exactly defining the content and scope of an individual right to the environment'[63] in the light of the fundamental importance of ecological systems. To this end, the report identifies a number of limiting factors to a human right such as the discrepancy between individual risks and collective risks,[64] the reductionist concept of state sovereignty, and restrictions resulting from the 'individualization' of the environment including the competition between individual rights and the difficulty of anthropocentric rights in embracing ecocentrism.[65] All these factors make alternative approaches more attractive including constitutional state obligations or separate rights for animals and plants, although the report dismisses this latter option as impractical.[66] State obligations to protect the environment are, in fact, favoured in

---

58   Resolution on a draft Constitution of the European Union 1994, OJC 61/155.

59   Ibid., Title VIII (21): 'Everyone has the right to the protection and conservation of his natural environment'.

60   Agudo, C. (2003), *Environment and Human Rights Report*, Committee on the Environment, Agriculture and Local and Regional Affairs, (16 April 2003; Doc. 9791), available at <http://assembly.coe.int/Documents/WorkingDocs/doc03/EDOC9791.htm>, accessed 1 December 2007.

61   Ibid., Explanatory Memorandum, 1.2.

62   Ibid.

63   Ibid. 2.4.

64   Ibid, 2.5.3.

65   Ibid. 2.5.4.

66   Ibid.

a number of European countries,[67] while an equal number of European countries has opted for an individual right to a healthy environment.[68]

All things considered, the report recommends a right to a healthy environment as 'a logical extension of [the European Court of Human Right's] present case law', but adding that 'it will not, of course, be possible to solve all the problems of environmental law'.[69]

Reviewing the case law generated by the European Court of Human Rights, we can see that the right to a healthy environment has, to some degree, been recognized. In *Taskin and Others v. Turkey*, mentioned earlier, the Court refers to Article 8 of the Turkish constitution and notes that a domestic court had squashed the operating permit 'based ... on the applicants' effective enjoyment of the right to life and the right to a healthy environment. In view of that conclusion, no other examination of the material aspect of the case with regard to the margin of appreciation is necessary'.[70] This and similar references are indications of the Court's validation of a 'right to live in a healthy and balanced environment'.[71] With respect to 'the relevant texts on the right to a healthy environment',[72] the Court then considers environmental rights contained in the Rio Declaration[73] and the Aarhus Convention,[74] as well as the mentioned recommendation by the Parliamentary Assembly of the Council of Europe. Although the Court stopped short of commenting on the current status of the right to a healthy environment, its reflections are significant as they were not necessary given the Court's ultimate reliance on domestic law. Overall, the European Court has been sympathetic to the idea of a human right to a healthy environment.

The Current Picture

Considering the development of case law and Community law as well as the constitutions in 13 EU member states[75] and a further six European states,[76] we can

---

67 Including the constitutions of Austria (Articles 10–12), Finland (Article 20), Germany (Article 20, a), Greece (Article 24), Netherlands (Article 21), Sweden (Article 2-2) and Switzerland (Article 24-7).

68 In the constitutions of Belgium (Article 23-4), Hungary, (Ch. I, Sec. 18), Norway (Article 110, b), Poland (Article 71), Portugal (Article 66-2), Slovakia (Articles 44 and 45), Slovenia (Articles 72, 73), Spain (Article 45-1) and Turkey (Article 56).

69 Environment and Human Rights Report (n. 60 above), 2.5.

70 *Taskin and Others v. Turkey* (n. 41 above), para. 117.

71 Ibid. paras. 131–2.

72 Ibid., Part II, B., paras. 98 et seq.

73 Ibid. para. 98.

74 Ibid. para. 99.

75 In addition to the constitutions of Belgium, Bulgaria, Czech Republic, Finland, Hungary, Latvia, Norway, Poland, Portugal, Slovakia, Slovenia and Spain, France has recently adopted the 'Environment Charter' ('Charte de l'environnement') declaring the 'right to live in a balanced and health-respecting environment' ('Chacun a le droit de vivre dans un environnement équilibré et respectueux de la santé'.), available at <http://www.ecologie.gouv.fr/IMG/pdf/affiche_charte_environnement.pdf>, accessed 1 December 2007.

76 Croatia, Macedonia, Russia, Ukraine, Moldova and Turkey.

conclude that Europe and the EU, in particular, have widely embraced the idea of a human right to a healthy environment. At the global scale, 56 constitutions have explicitly recognized the right to a clean and healthy environment.[77] This wide recognition, together with international soft law development, would suggest that the right to a healthy environment is a human right in statu nascendi.

On the other hand, 97 constitutions have gone the other way, at least, for the time being. They contain provisions that make it the duty of the national government to prevent harm to the environment.[78] The reasons for favouring state obligations over a rights approach may differ among countries, however, it is significant that there is no uniform response to conceptualizing environmental rights and responsibilities. Some prefer the rights of citizens, others prefer obligations of the government, and some countries provide combinations of both.[79] Furthermore, 56 constitutions recognize a responsibility of citizens or residents to protect the environment.[80]

Even within the system of environmental rights there is limited conformity. It is clear that procedural environmental rights are firmly established in soft and hard law. Most countries have provisions to allow, in some form, access to environmental information, decision-making and judicial review. With respect to substantial rights there is recognition, on the one hand, of the environmental dimension of existing human rights and, on the other hand, of a distinct human right to a healthy environment.

In terms of real outcomes, a distinct human right does go further; however, there is uncertainty about its content. Dinah Shelton has argued that the right to environment includes elements of intergenerational equity and aesthetic protection.[81] Lynda Collins has recently argued that it should also be understood to include the precautionary principle.[82] These and possibly more aspects of ecological sustainability could be part of the content. But are they? There is little evidence that courts went much beyond what is already being granted by the human right to life, well-being, privacy, property, and so on. It appears that the anthropocentric nature of human rights does not allow for ecocentrism nor, therefore, for a recognition of ecological sustainability as a content that would distinguish a right to a healthy environment.

In conclusion, human rights and the environment are inextricably linked to each other. Without human rights, environmental protection could not effectively be enforced. And vice versa, without the inclusion of the environment, human rights would be in danger of losing their core function, i.e. the protection of human life, well-being and integrity.

---

77   See Mollo, M. et al. (2005), *Environmental Human Rights Report: Human Rights and the Environment – Materials for the 61st Session of the United Nations Commission on Human Rights, Geneva, March 14–April 22, 2005* (Oakland, CA, Earthjustice Legal Defense Fund), 37 at footnote 172.

78   Ibid. 37.

79   The Mollo Report 2005 (n. 77 above) does not give the exact number of combined systems, but it can be estimated at about 30.

80   Ibid. 38.

81   Shelton, D. (1991), 'Human Rights, Environmental Rights, and the Right to Environment', *Stanford Journal of International Law* 28, 103–38, 133–34.

82   Collins 2007 (n. 20 above).

The overview of international and European law with respect to environmental human rights shows some trends. There is increasing legal recognition of the idea that environmental degradation can result in deprivations of existing human rights. There is also increased awareness that mere recognition of such deprivations is not enough to promote and secure a healthy environment. To achieve this, two approaches have been followed: one is to strengthen procedural environmental rights, the other is to recognize a distinct human right to a healthy environment. Clearly, international and national environmental law have embraced the idea that traditional human rights concepts are insufficient to accommodate concerns for environmental protection and sustainability.

However, will the addition of procedural rights and a right to a healthy environment to the catalogue of human rights be sufficient? Or is there a need to revisit the core idea of human rights understood as protection from arbitrariness and misuse of power? Could it be that human beings need protection 'from themselves'? If the notion of 'humans as the most dangerous animal on Earth' holds some truth, then the issue of environmental duties to complement environmental rights arises. This issue has been widely overlooked in the environmental rights debate. There is as yet no theory on how environmental rights may be related to environmental duties. While the former appear as legal entitlements, the latter are, at best, referred to as moral obligations. From an ecological perspective, such imbalance reveals utilitarianism and anthropocentrism. Imposing mere moral duties upon ourselves is certainly insufficient to recognize ecological sustainability of all life in a legal sense. The ecological approach to human rights holds that not only humans, but also non-human beings are entitled to the protection of their life, well-being and integrity, albeit not necessarily in the same way. Human rights operate not only in a social, but also in an ecological context. This reality has yet to be reflected in human rights theory and practice.

Sustainability Ethics and Human Rights

There is concern amongst many commentators over the inherent anthropocentricity of environmental human rights. In the view of some, their very existence reinforces the idea that the environment exists only for human benefit and has no intrinsic worth. Furthermore, they result in creating a hierarchy, according to which humanity is given a position of superiority and importance above and separate from other members of the natural community.[83] More specifically, the objectives and standards applied are human-centred. Humanity's survival, living standards, and continued use of resources are the objectives. The state of the environment is determined by the needs of humanity, not the needs of other species.

This human-centred character of an environmental human right leads to a philosophical tension between deep (ecocentric) and shallow (anthropocentric)

---

83 Birnie and Boyle 2002 (n. 6 above), 257–8.

ecologists. As a result of this tension, some commentators wholly reject human rights proposals,[84] while others offer a compromise position.[85]

Those who condemn the human rights approach raise the following concerns. First, anthropocentric approaches to environmental protection are seen as perpetuating the values and attitudes that are at the root of environmental degradation. Second, anthropocentric approaches deprive the environment of direct and comprehensive protection. For example, 'human' life, health, and standards of living are likely to be the aims of environmental protection. Thus the environment is only protected as a consequence of, and to the extent needed to protect human well-being. An environmental right thus subjugates all other needs, interests and values of nature to those of humanity. Environmental degradation as such is not sufficient cause for complaint, it must be linked to human well-being. Third, humans are the beneficiaries of any relief for infringement of the right. There is no guarantee of its utilization for the benefit of the environment. Nor is there any recognition of nature as the victim of degradation. Fourth, environmental protection is dependent on human protest.

On the other hand, a number of arguments are put forward which may, to some extent, mitigate these concerns. First, it is suggested that a degree of anthropocentrism is a necessary part of environmental protection. Not in the sense of humanity as the centre of the biosphere, but because humanity is the only species, that we know of, which has the consciousness to recognize and respect the morality of rights and because human beings are themselves an integral part of nature. In short, the interests and duties of humanity are inseparable from environmental protection.

Thus far, Shelton agrees,[86] but goes on to argue that an environmental human right could be complementary to wider protection of the biosphere which recognizes the intrinsic values of nature, independent of human needs. Birnie and Boyle, on the other hand, point out that 'by looking at the problem [of anthropocentric human right] in moral isolation from other species such a right may reinforce the assumption that the environment and its natural resources exist only for human benefit, and have no intrinsic worth in themselves'.[87] They see the implications of the issue as being largely structural, requiring the integration of human rights claims within a broader decision-making framework capable of taking into account, amongst other factors, intrinsic values, the needs of future generations and the competing interests of states. In their view, human rights institutions are currently too limited in their perspective to be able to balance these factors.[88]

Rolston also advocates a compromise position. He accepts the paradigm of human rights for the protection of human needs for environmental integrity, but in

---

84   Gibson, N. (1990), 'The Right to a Clean Environment', *Saskatchewan Law Review* 54, 5–7; Giagnocavo, C. and Goldstein, H. (1989–1990), 'Law Reform or World Re-form', *McGill Law Journal* 35, 345, 346.

85   See generally Kiss and Shelton 2004 (n. 11 above); and Nickel, J.W. (1993), 'The Human Right to a Safe Environment: Philosophical Perspectives on its Scope and Justification', *Yale Journal of International Law* 18, 281–95.

86   Shelton 1991 (n. 81 above), 110.

87   Birnie and Boyle 2002 (n. 14 above), 257.

88   Ibid.

addition suggests the elaboration of human responsibilities for nature.[89] According to Nickel, human rights play a 'useful and justifiable role in protecting human interests in a safe environment and in providing a link between the environment and human rights movements'.[90] He labels his approach as 'accommodationist', arguing that anthropocentrism is not a significant objection if 'it can be supplemented by other norms that will address other issues'.[91] In other words, it could be seen as a useful part of 'the normative repertory of environmentalism'.[92]

In the short term, these approaches might be useful in assisting environmental law to transform from an essentially anthropocentric perspective to an ecocentric perspective. However, in the long term the existence of an environmental human right could be seen as self-contradictory. A better option is the development of all human rights in a manner which demonstrates that humanity is an integral part of the biosphere, that nature has an intrinsic value and that humanity has obligations towards nature. In short, ecological limitations, together with corollary obligations, should be part of the rights discourse.

Attempts to overcome the anthropocentric approach are plentiful. Among these, the concept of nature's rights has been well documented since its rise to prominence in 1972, following the publication of Christopher Stone's article 'Should Trees Have Standing?'[93] Over more than 30 years the concept has been debated amongst lawyers, philosophers, theologians and sociologists. This debate has lead to an advocacy of a wide variety of rights approaches including: legally enforceable rights for nature (as envisaged by Stone); so-called 'biotic rights' (being moral imperatives which are not legally enforceable); moral 'responsibilities'; and 'rightness' (a norm which prescribes a need for a proper healthy relationship between humanity and nature). What is common to each is an attempt to give concrete and meaningful recognition to the intrinsic value of nature. They differ in how this should be achieved. Some commentators advocate that it should be done within the context of legally enforceable rights, others argue for recognition through enunciation of values or status, which requires humanity to take into account the interests of nature and to accord these interests a priority that they might not otherwise be granted.

Giagnocavo and Goldstein have argued that the concept of nature's rights is tantamount to a 'quick legal fix', which, like many other legal solutions, precludes the deep questions necessary for genuine world change.[94] In particular, they have challenged the theory that 'rights' are an appropriate method of social reform, leading us to change our attitudes and value entities (in this case nature) to which 'rights'

---

89 Rolston, H. III (1993), 'Rights and Responsibilities on the Home Planet', *Yale Journal of International Law* 18, 251–79, 259–62.

90 Nickel (n. 85 above), 282.

91 Ibid. 283.

92 Ibid.

93 Stone, C. (1972), 'Should Trees have Standing?', *Southern California Law Review* 45, 450–501. See also Stone, C. (1987), *Earth and Other Ethics: The Case for Moral Pluralism* (New York, Harper & Row) and Stone, C. (1996), *Should Trees Have Standing? And other Essays on Law, Morals and the Environment: 25th Anniversary Edition* (New York, Oxford University Press).

94 Giagnocavo and Goldstein 1989–1990 (n. 84 above).

are ascribed. Giagnocavo and Goldstein reject this theory as a 'false claim'. In their opinion, legal 'rights' give the holder some advantages (as discussed by Stone), but this only amounts to valuing by legal institutions, not society at large.[95]

Stone himself recognizes the limitations of his 'rights' theory and in the final pages of his article discusses the importance of a changed environmental consciousness. He states that legal reform, together with attendant social reform, will be insufficient without 'a radical shift in our feelings about "our" place in the rest of Nature'.[96] Stone has never considered 'rights' as an end in themselves but rather as a means to an end.[97]

It will be some time before we see a major international treaty reflecting a legal position beyond anthropocentrism. The fate of the United Nations Earth Charter perhaps best illustrates this. The Earth Charter document was to be a 'short, uplifting inspirational, and timeless expression of a bold new global ethic'.[98] However, as the negotiating process dragged on it ended up being called the 'Rio Declaration', which many criticize as being little more than declaratory of the social and political conflicts which infused all UNCED negotiations. NGOs took up the challenge when UNCED was seen to fail in its objective and drafted their own Earth Charter. The NGO Earth Charter does not shy from the task of accepting responsibility for nature and defines it in ecocentric terms.[99] The Preamble states 'We accept a shared responsibility to protect and restore Earth and to allow wise and equitable use of resources so as to achieve an ecological balance and new social, economic and spiritual values'.[100]

In the absence of a clear statement of a new ethic, in an international document such as the proposed UN Earth Charter, developments in the area of environmental human rights may produce some flow-on effects for the creation of nature's rights. In considering constitutional entrenchment of an environmental right, those states most concerned to avoid the anthropocentricity of such rights are likely to explore, in some detail, the notion of nature's rights.[101] In addition, municipal developments which take into account the differing moral and legal traditions of indigenous peoples, as

---

95   Ibid. 357.

96   Stone 1972 (n. 93 above), 495. See generally 489–501.

97   See also Bosselmann, K. (1993), Introduction to Stone, C., *Umwelt vor Gericht* (German tr. of *Should Trees have Standing?*), 2nd edn (Munich, Trickster Verlag).

98   Grubb, M. et al. (1993), *The Earth Summit Agreements: A Guide and Assessment* (London, Earthscan), 83.

99   As a negotiator of the NGO Earth Charter I was surprised to see an undisputed commitment to ethical ecocentrism among the hundred or so NGOs represented at these negotiations in Rio.

100   The Draft Earth Charter is published in Pacific Institute of Resource Management (ed.) (1992), 'Commitment for the Future: The Earth Charter and Treaties agreed to by the International NGOs and Social Movements', Paper presented to the International NGO and Social Movements Forum Conference, Wellington, New Zealand, 11 June. The first principle states: 'We agree to respect, encourage, protect and restore Earth's ecosystems to ensure biological and cultural diversity'.

101   See Stone's discussion of the *Seehunde v. Bundesrepublik Deutschland* (in which an action was taken on behalf of seals suffering from chemical pollution of the Baltic Sea) in Bosselmann, K. (1992), *Im Namen der Natur – Der Weg zum ökologischen Rechtsstaat*

these traditions apply to nature, may also have some influence on the international legal system. The international legal system, like municipal systems, is becoming increasingly cognizant of the wisdom of indigenous cultures.

Whether or not the concept of nature's rights is ever implemented by international or municipal law, the very existence of the debate contributes to the development of ecological rights. It helps develop consciousness beyond the prevalent anthropocentric ethic by suggesting what to many might have formerly been the 'unthinkable'.[102] Gradual acceptance of moral responsibilities towards nature may lead to a point where we begin to accept the idea of ecological limitations on the exercise of our rights or, more directly, agree to redefinition of the content of certain rights (e.g. property rights).[103] On the other hand, the limitations of the nature's rights debate must also be borne in mind. We must guard against its overextension. This can be achieved by seeing nature's rights within an appropriate context. Thus when legal processes are involved we must acknowledge the limitations of these processes. However, these limitations do not necessarily deprive rights of their usefulness as a tool in the process of transition. Their use, in conjunction with other changes in society, may result in the creation of certain resonances in all social systems which will in turn lead to enduring change. This context also includes recognition that other important parallel changes must also occur, for example, a change in consciousness. Law and pleas for new morality cannot and do not exist in vacuums, nor can we rely on them to provide solutions to our deepest and most complex problems.

The Ecological Approach to Human Rights

Many environmental lawyers have questioned the fundamentally anthropocentric character of environmental law. They are calling for an ecocentric turnaround. Some have argued, therefore, that we should not view environmental issues through a human rights focus, entailing a form of 'species chauvinism' (D'Amato 1990). We should instead think either of nature's rights or of limitations of human rights with respect to the 'intrinsic values' of the environment.

The former idea of rights for nature has been described as the 'strong rights-based approach', the latter idea of intrinsic values as the 'weak rights-based approach'[104] which is what is advocated here. There is little cause to believe that an ecocentric turnaround can be achieved just by adding rights of nature to the catalogue of the rights of humans. As seen above, there are a number of difficulties with 'rights' thinking, the most important being that we would only foster the anthropocentric and

---

(Munich, Scherz), 181–9; and Stone, C. (1993), *The Gnat is Older than Man: Global Environment and Human Agenda* (Princeton, Princeton University Press), 85–6.

102   Stone 1972 (n. 93 above), 453–7.

103   Nash, J.A. (1993), 'The Case for Biotic Rights', *Yale Journal of International Law* 18, 235–49.

104   Redgwell, C. (1996), 'Life, the Universe and Everything: A Critique of Anthropocentric Rights', in Boyle, A. and Anderson, M. (eds), *Human Rights Approaches to Environmental Protection* (Oxford, Clarendon Press), 71, 73.

individualistic tradition of rights, which represents the very mind set that has caused the global environmental crisis in the first place.

The concept of ecological human rights[105] attempts to reconcile the philosophical foundations of human rights with ecological principles. The aim is to link the intrinsic values of the humans with the intrinsic values of other species and the environment. As a result, human rights (such as human dignity, liberty, property, development) need to respond to the fact that the individual not only operates in a social environment, but also in a natural environment. Just as much as the individual has to respect the intrinsic value of fellow human beings, the individual also has to respect the intrinsic value of other fellow beings (animals, plants, ecosystems).

The reference to 'respect' for others as the determining factor for individual freedom is not incidental. In both the literature on environmental ethics and the literature on human rights there is certain common ground. Ethical considerations on our relationship with the environment often use the category of respect as, for instance, Paul Taylor in his influential work Respect for Nature (1986) or Tom Regan in his discussion of moral and legal obligations.[106] The contemporary ethical debate is largely focused on intrinsic values as the basis for moral consideration and respect as the basis for personal obligations.[107] Much of the impetus for respect probably lies with Kant, whose insistence that persons should be treated as ends and not only means has become the touchstone of modern day humanism. While Kant's focus is on respect for persons rather than respect for life, this need not necessarily exclude life as the proper object of respect; the extension of the persona concept to include non-human entities has been advocated by environmental lawyers including Christopher Stone and Tom Regan. However, from a Kantian perspective such extension is prohibitive as it is closely bound up with the perception of persons as choosers or centres of rational consciousness, and 'life' does not lend itself so easily to such attributions. But this does not mean that talk of respect for (the intrinsic value

---

105   Bosselmann 1995 (n. 4 above); see further Bosselmann, K. (2000), 'Un Approcio Ecologico Ai Diritti Umani' (An Ecological Approach to Human Rights), in Greco, M. (ed.), *Diritti Umani E Ambiente (Human Rights and the Environment)* (Fiersole, Edizioni Cultura della Pace), 67–87; Bosselmann, K. (2001a), 'Human Rights and the Environment: The Search for Common Ground', *Revista de Direito Ambiental*, 23 July–September, 12–28; see further, Bosselmann 1992 (n. 102 above), 181–249; Bosselmann 1995 (n. 4 above), 222–63; Bosselmann, K. (2001b), *Ökologische Grundrechte* (Baden-Baden, Nomos); Bosselmann, K. and Schröter (2001), *Umwelt und Gerechtigkeit* (Baden-Baden, Nomos); Taylor, P (1998a), *An Ecological Approach to International Law* (London, Routledge), 196–257; Taylor, P. (1998b), 'From Environmental to Ecological Rights: A New Dynamic in International Law?', *Georgetown International Environmental Law Review* 10:2, 309–97; Taylor, P. (2008), 'Ecological Integrity and Human Rights', in Westra, L., Bosselmann, K. and Westra, R. (eds), *Reconciling Human Existence and Ecological Integrity* (London, Earthscan).

106   Regan, T. (1992), 'Does Environmental Ethics Rest on a Mistake?', *Monist* 75, 161–82.

107   See Elliot, R. (ed.) (1996), *Environmental Ethics* (Oxford, Oxford University Press), 15.

of) life makes no sense. Respect in the sense of recognition and referentia (rather than Kant's observantia) is not limited to the rational consciousness.[108]

In human rights theory we often find the concept of 'respect' as the basis of human rights. Again, Kant's category of respect has been influential, and limitations of this category to a non-anthropocentric concept of human rights would apply. Nevertheless, the inclusion of non-human entities can be achieved. McDougal, Lasswell and Chen, in their standard text on human rights (1980), for example, suggest that using respect as an universal principle would allow the inclusion of all aspects of life in the protection of fundamental rights. John Rawls's Theory of Justice may not be far from this with its emphasis on a universal principle which needs to be accepted by all in order to create a just society.[109] The respect for the intrinsic value of life could guide both the relationship between the individual and society on the one hand and the relationship between humans and the environment on the other.

The Social Dimension of Human Rights

Structurally, human rights can be limited by ecological considerations in the same way as they are presently limited, namely by social and democratic considerations. Human rights are not absolute, but subject to a variety of limiting factors. There are general and specific limitations to individual rights. A whole variety of limitations exists in the human rights catalogue of the German constitution as the following extracts illustrate:

Article 1 – Protection of Human Dignity:

(1) The dignity of the human being is inviolable. ...
(2) The German people acknowledges inviolable and inalienable human rights as the basis of every community, of peace and justice in the world.
(3) The following basic rights shall bind the legislature, the executive and the judiciary as directly enforceable law.

Article 2 – Right of Liberty:

(1) Everyone shall have the right to the free development of his/her personality in so far as he/she does not violate the rights of others or offend against the constitutional order or the moral code.

...

---

108   Kleinig, J. (1991), *Valuing Life* (Princeton, Princeton University Press), 18.

109   The individualistic approach of Rawls's liberalism is, of course, a different matter; for a general critique see e.g. Douzinas, C. and Warrington, R. (1994), *Justice Miscarried* (New York and London, Harvester Wheatsheaf); Bosselmann, K. (1998), 'Justice and Environment: Building Blocks for a Theory of Ecological Justice', in Bosselmann, K. and Richardson, B. (eds), *Environmental Justice and Market Mechanisms* (London, Kluwer), 30–57; Bosselmann, K. (2006), 'Ecological Justice and Law', in Richardson, B. and Wood, S. (eds), *Environmental Law for Sustainability: A Critical Reader* (Oxford, Hart), 129–63.

Article 5 – Freedom of Expression:

...

(3) Art and science, research and teaching shall be free. Freedom of teaching shall not absolve from loyalty to the constitution.

Article 14 – Property:

(1) Property and the right of inheritance are guaranteed. Their content and limits shall be determined by the laws.
(2) Property imposes duties. Its use should also serve the public well-being.

A general reference often used in legislation is the 'reasonable limits prescribed by law as can be demonstrably justified in a free and democratic society'. This phrase appears, for instance, in the European Convention for the Protection of Human Rights and Fundamental Freedoms, in the Canadian Charter of Rights and Freedoms and in the New Zealand Bill of Rights. Typically, any limitation to an individual right has to pass proportionality tests of necessity, lowest possible impairment and balance of conflicting rights.

There is considerable variation as to how the balancing is actually achieved. For example, civil law countries and the United States follow an 'absolutist approach', under which there is heavy emphasis on the supremacy of law, particularly the Constitution, and an attempt to avoid substantive issues. On the other hand, countries like Britain, Australia or New Zealand follow a 'balancing of interests' approach which attempts to weigh up the various interests. However, the bottom line is the same throughout all these jurisdictions. It is always the concern for the rights of all members of society which ultimately determines to what extent the rights of the individual may be limited. This bottom line can be referred to as the 'social dimension of human rights'.[110]

The essence of human rights appears to be the attempt to define the freedom of the individual in interaction with other individuals. Thus, it is the social sphere of human existence which human rights are concerned with, not the biosphere. The biosphere (environment) is presently taken for granted and has no legal quality. Human rights are historically and systematically created to protect citizens against the state, in other words to protect humans from each other; they contain no provision to stop humans from exploiting non-humans and fundamentally changing the conditions of life. As long as human rights are not impinged on we are free to destroy the environment and all life around us.

The only existing restriction in this respect is our anthropocentric morality which may require us not to torture animals, not to turn a beautiful landscape into a moonscape or to limit genetic engineering to those areas beneficial to we humans.

---

110 Which is the common term used in German theories on fundamental rights. Accordingly, an ecological dimension of human rights can be postulated; see Bosselmann, K. (2001), 'Human Rights and the Environment: Redefining Fundamental Principles?', in Gleeson, B. and Low, N. (eds), *Governing for the Environment* (Basingstoke, Palgrave), 132–4.

The limits are always drawn by our concern for human welfare to the exclusion of the welfare of other life forms. The dilemma is, of course, that we cannot survive without a concern for the welfare of life as a whole. This is the harsh reality discovered by ecology.

Anthropocentric limitations of Western tradition have a long history and are perhaps systemic. However, what makes them so dangerous today, literally life-threatening, is that they reinforce human arrogance even in the most advanced environmental legislation. The law cements the view that only humans matter and the environment has just instrumental value – a view with grave ecological blindness.

To overcome this blindness there two options. Either we manage the ethical paradigm shift in society and don't worry about human rights; we may simply assume that human insight will prevail. Or we promote the ethical paradigm shift at all social levels including the law.

Without discussing to what extent the law can make a difference to social behaviour, both of the two classic views appear to be wrong. Neither the traditional liberal view, which holds that there is a profound difference between legal norms and social reality, is true, nor is the Marxist view, which denies any difference between legal norms and social reality, appropriate. The law is both purely reflecting and actively influencing the way in which society operates. That is why it matters whether ecological reflections exist in legal norms or not.

The Ecological Dimension of Human Rights

For a concept as revolutionary as the non-anthropocentric concept of human rights, the burden of proof is, if course, with those advocating it. What then is the advantage of ecological human rights? Would they make any difference to the real outcome of decision-making? One example should illustrate this. It will demonstrate why it would not be sufficient to purely rely on the social dimension of human rights.

The example is the law related to biotechnology. At international level, biotechnology became the subject of international law through the 1992 Convention on Biological Diversity.[111] Along with a general trend in recent international environmental law, the Biodiversity Convention takes the approach of ecosystem protection (i.e. protecting entire habitats rather than individual species as such).[112] It does so by introducing (in its Preamble) an 'intrinsic value of biological diversity', in addition to 'the ecological, genetic, social, economic, scientific, educational, cultural, recreational and aesthetic values of biological diversity and its components'. This is the recognition of the distinction between (ecocentric) intrinsic values and (anthropocentric) instrumental values of the environment.

In fact, there is a distinct body of environmental agreements with an ecocentric focus.[113] Examples include the 1991 Protocol on Environmental Protection amending the 1959 Antarctic Treaty, the 1982 World Charter for Nature, and the 32 so-called

---

111   Convention on Biological Diversity (Rio de Janeiro, 3–14 June 1992; UN Doc. 6.10, 31 ILM 818 (1992)).

112   Kiss and Shelton 2004 (n. 11 above), 17, 288–382.

113   Redgwell 1996 (n. 104 above), 71–87.

Alternative Treaties which several hundred non-governmental organizations negotiated at the 1992 Earth Summit in Rio. Article 4 of the 1995 Draft International Covenant on Environment and Development establishes the principle of respect for all forms of life and Principle 1 of the 2000 Earth Charter sets out 'respect Earth and life in all its diversity'.

Article 19 of the Biodiversity Convention calls for the contracting states to take legislative measures towards controlling biotechnological research activities. The problem is that the Convention, like most treaties, leaves the means of implementation totally to the discretion of states.

At municipal level, several countries have introduced such controlling legislation, among them Germany with its Gentechnikgesetz (Gene Technology Law) of 1990. Such legislation regulates details of notification and licensing of genetically modified products (such as, for example, the release of those products into the environment), but it always does so on the basis that there is a fundamental right to conduct genetic engineering in the first place. The principle of free production and sale is the rule, any restrictions are the exception. The burden of proof, therefore, is not with the producer introducing a new risk potential, but with the general public (represented e.g. by expert commissions such as the Environmental Risk Management Authority in New Zealand or various commissions in the United Kingdom). Whether or not activities of genetic engineering are acceptable is determined by weighing up social costs and benefits. The problem is that such social costs and benefits are exclusively determined by values of human utility. There are no intrinsic values of ecosystems and their components to be considered.

Quite obviously, there is a gap between the ecocentric approach of the Biodiversity Convention and its implementation through the anthropocentric approach of municipal legislation. To close this gap, one could imagine a simple legislative act to impose the burden of proof on the producer (or importer) with the consequence that any remaining doubts go against the applicant. However, such a radical interpretation of the polluter pays and precautionary principles has not been made anywhere and it is unlikely, indeed impossible, for it to be made on the basis of our current anthropocentric concept of human rights.

Research, development and commercial application of genetic engineering are considered free up to a point where the rights of others may be impinged on. Such affected rights may include consumer rights (like the right to make informed choices), rights of health protection (i.e. against human health risks associated with genetically modified products), and perhaps human dignity or the right to personal identity and self-determination. However, once these concerns are met nothing could stop genetic engineering from fundamentally altering the genetic structures of which nature is made up. That is why, for example, cloning of humans may be seen as restricted by the principle of human dignity or the right to personal identity and self-determination, but cloning of animals and plants is not. This would be purely an issue of utilitarian considerations. If the 'Dolly experiments' appear useful to humans and their immediate needs, they will be considered lawful.[114] Sheep, like all animals and plants, are at the receiving end of our anthropocentric morality.[115]

It may be, of course, that our morality will change over time and that, one day, ethical committees will have the wisdom and power to stop genetic engineering going mad. At the moment, ethical committees are guided by absolute freedom of research on the one hand and utilitarian cost–benefit analysis on the other. Since both principles are firmly enshrined in our human rights concept, the long-term ecological implications of genetic engineering will not count.

A closer examination of current case law reveals that ecological human rights would have altered the outcome. For instance, with respect to property rights German courts have increasingly acknowledged that land and resource use is restricted by requirements of the 'public weal' (Article 14). This led, for example, to restrictions in the use of chemical fertilizers and pesticides on farmland, to protection against overgrazing caused by too many cattle and a ban on certain hazardous substances. However, in all cases the restrictions were ultimately determined by human health standards, not ecological concerns. As the German Federal Constitutional Court (in a case of 1982 regarding ground water levels) stated: 'Private land use is limited by the rights and interests of the general public, to have access to certain assets essential for human well-being such as water'.[116] Respect for the intrinsic value of life (other than human life) would have led to much more stringent restrictions than securing water supply for people. However, to quote from another decision, this time of the German Federal Administrative Court (1987): 'The law cannot provide for the health of ecosystems *per se*, but only in so far as required to protect the rights of affected people'.[117]

A notable exception to this anthropocentric reductionism (and proving the rule) is the protection of animals. Various European countries have, in the past few years, altered the legal status of animals. They are not regarded any more as 'things', which can be possessed and used like cars, but as 'creatures' in their own right. Consequently, there are now a number of cases penalizing 'inhumane' treatment of animals (e.g. banning certain means of killing them or requiring a minimum cage size for poultry). So, the recognition of, at least, a rudimentary form of the intrinsic value of animals has made a significant difference. Apparently, the animal rights movement of the 1970s and 1980s is bearing fruit.

Some legal commentators now speak of a remarkable 'spill-over effect' caused by the international trend towards a human right to a decent environment and towards recognition of animal rights. Both clearly anthropocentric, in nature, the 'spill-over effect' is, nevertheless, clearly there. To quote Catherine Redgwell, 'The dam of anthropocentrism has clearly been breached. Given the increasing awareness of the interconnectedness of human beings and the environment and of the intrinsic value

---

114 Which, of course, they are at present.

115 For a critique see Lowry, M.L. (1996), *Of Mice and Genes: Ethics and European Patent Law on Biotechnological Inventions* (European University Institute); Bosselmann, K. (1998), *Ökologische Grundrechte* (Baden-Baden, Nomos).

116 Bosselmann 2001 (n. 110 above), 94–7.

117 Ibid.

of the latter ... nature is unlikely to simply be ignored; rather, the problem is one of reconciling a diverse environmental (agenda) and human rights agenda'.[118]

The German Constitutional Debate since 1985

The reconciliation of these two agendas can be achieved by integrating the environment into the concept of human rights. Human rights can be shaped by limitations drawn from both their social and ecological context. In Germany, such consideration has been part of a wider debate surrounding core values of the Grundgesetz. One of these core values concerns the shift to a non-anthropocentric approach.

The constitutional development in Germany reflects both support for and opposition to this idea. In the mid 1980s the penetrating powers of ecologism were strong enough to instigate a broad public debate on the merits of a new state objective. State objectives ('Staatsziele') are binding constitutional law requiring government to seek to fulfil certain tasks. The incorporation of a state objective to protect the environment quickly found acceptance, but its reason, purpose and extent were highly controversial, eventually revealing two blocs. One bloc demanded the state to protect the environment for its own sake ('um ihrer selbst Willen'), the other insisted on the environment as mere natural resources of human beings ('natürliche Lebensgrundlagen des Menschen').[119] The former bloc consisted of environmental groups, environmental lawyers, churches and the then political opposition of Social Democrats and Greens, the other bloc was made up of constitutional lawyers and the government. The Joined Constitutional Commission of the Federal Parliament ('Bundestag') and the Federal Senate ('Bunderat') found the question of anthropocentricity or non-anthropocentricity too important to be decided at this stage and called for more public debate. The Commission concluded, however, that the environment as such could not carry a similar constitutional intrinsic value comparable with human beings.[120]

The introduction of a new Article 20a in 1994 represented a political compromise between both blocs: 'The state, also in its responsibility for future generations, protects the natural foundations of life in the framework of the constitutional order, by legislation, and, according to law and justice, through the executive and the courts'.

The introduction of 'natural foundations of life' (rather than human life) marked a step away from crude anthropocentricity. But the debate did not stop in 1994. Now the animal rights movement lobbied for a specific state objective to protect animals. In 2002, the phrase 'and the animals' was added to the 'natural foundations of life'. Whether this addition has strengthened or weakened a non-anthropocentric approach to the constitution may be a matter of interpretation. It does prove, however, that the ethical discourse has made inroads into the legal discourse.

Constitutional changes have been even more successful in Switzerland. A 1992 amendment to the Federal Constitution required the state to take into account the

---

118   Redgwell 1996 (n. 104 above), 73.
119   See Bosselmann 1995 (n. 4 above), 195–202.
120   Bosselmann, K. 1998 (n. 115 above), 80–82.

Würde der Kreatur.[121] This notion might be translated as 'dignity of creation',[122] except this resembles the German term Schöpfung (reflecting Christian terminology). The other official language, French, captures the idea of *Würde der Kreatur* much better: 'l'intégrité des organismes vivants'. This 'integrity of living organisms' would be very close to ecological integrity as, for example, expressed in the Earth Charter. However, there is ongoing dispute about the true meaning of *Würde der Kreatur*.

According to Peter Saladin *Würde der Kreatur* has an essential core that must not be infringed upon and cannot be set aside by a balancing process. In his 1994 report for the Swiss Environmental Protection Agency (EPA) he emphasized that Würde des Menschen and *Würde der Kreatur* do not point to something substantively different.[123] Both reflect intrinsic value and dignity. Not surprisingly, the EPA ordered a second opinion. It came from a Zurich University ethics group arguing that *Würde der Kreatur* is meant to be seen at a different level from human dignity. The group's 1997 report[124] called for the more narrow interpretation of dignity that protects individuals from degradation. Full recognition of ecological integrity remains on the agenda and only time will tell whether constitutions will incorporate ecocentrism.

With respect to human rights, the German constitutional reform saw various attempts to formulate ecological limitations. A proposal by the state of Bremen included a state obligation to protect the natural world ('natürliche Mitwelt') for its own sake ('um ihrer selbst willen') and an ecological restriction of individual freedoms, for example, in Articles 2 (Right of Liberty) and 14 (Property). In the Upper House (Bundesrat), ten states voted for this proposal; six abstained.[125]

Following Germany's unification in 1990, a broad alliance of political scientists, constitutional lawyers and political parties drafted a new constitution ('Verfassung' to replace the 'Grundgesetz'). This draft constitution – the only one to date – made ecology a fundamental principle next to democracy, individual freedom and social justice.[126] Ecological responsibility is seen as a 'green thread throughout the entire

---

121 The new Swiss constitution of 2000 incorporates the equivalent 1992 article as Article 120 ('gene technology in the nonhuman area'):
1 Persons and their environment shall be protected against abuse of gene technology.
2 The Confederation shall legislate on the use of the reproductive and genetic material of animals, plants, and other organisms. In doing so, it shall take into account the *dignity of creation* and the security of man, animal and environment, and shall protect the genetic multiplicity of animal and vegetal species.
122 A translation offered by the Swiss government.
123 Praetorius, I. and Saladin, P. (1994), *Die Würde der Kreatur*, Schriftenreihe Umwelt Nr. 260, BUWAL (ed.), 121; see also the standard commentary on the Swiss constitution in Saladin, P. and Schweizer, R.J. (1995), *Kommentar zur Bundesverfassung der Schweizerischen Eidgenossenschaft*, Article 24, novies Abs. 3 (Basle, Verlag Helbing & Lichtenhahn).
124 Published in English as Balzer, P., Rippe, K.P. and Schaber, P. (2000), 'Two Concepts of Dignity for Humans and Non-Human Organisms in the Context of Genetic Engineering', *Journal of Agricultural and Environmental Ethics* 13, 7–27.
125 Bosselmann 1995 (n. 4 above), 200–202.
126 Kuratorium für einen demokratisch verfaßten Bund deutscher Länder (1991), *Vom Grundgesetz zur deutschen Verfassung. Denkschrift und Verfassungsentwurf* (Baden-Baden, Nomos), 21–3.

constitution'[127] affecting the state and the individual alike. The draft constitution rejects a human right to a healthy environment as it would reflect a 'problematic anthropocentric view point with respect to nature'.[128] Instead, several human rights contain limitations to reflect ecological responsibilities. The property concept of Article 14, for example, includes the preservation of natural conditions of life ('Erhaltung der natürlichen Lebensgrundlagen') as a barrier to property use.[129] In a similar way, freedom of science and research is restricted. Article 5 b(2) requires public notification for any research involving particular risks ('besondere Risiken') and allows restrictions if such research might cause a threat to human dignity or the natural conditions of life.[130]

The recognition of the intrinsic value of life is both the ethical and legal justification for ecological limitations. Some examples[131] – using the same German fundamental rights as shown above – can illustrate the use of this concept:[132]

Article 1 – Protection of Human Dignity:

(1) The dignity of the human being is inviolable. ...
(2) The German people acknowledges inviolable and inalienable human rights and the respect for the intrinsic value of life as the basis of every community, of peace and justice in the world.
(3) The following basic rights shall bind the legislature, the executive and the judiciary as directly enforceable law.

Article 2 – Right of Liberty:

(1) Everyone shall have the right to the free development of his/her personality in so far as he/she does not violate the rights of others or the sustainability of natural conditions of life.

...

Article 5 – Freedom of Expression:

...

(3) Art and science, research and teaching shall be free. They respect the dignity of the human being and the intrinsic value of life.

...

Article 14 – Property:

---

127   Ibid. 39.
128   Ibid. 40.
129   Ibid. 40, 86.
130   Ibid. 73.
131   Bosselmann 1995 (n. 4 above), 80–126.
132   Proposed amendments in italics.

(1) Property and the right of inheritance are guaranteed. Their content and limits shall be determined by the laws.
(2) Property imposes duties. Its use should also serve the public well-being and the sustainability of natural conditions of life.

The importance is not the exact wording,[133] but the intention or, more precisely, the dynamics carrying the ecological interpretation of human rights.

The Ecological Approach to Human Rights in the Earth Charter

An international example can be found in the Earth Charter.[134] As an ethical framework for a just, sustainable and peaceful future world, the Earth Charter sets out relevant values and principles including their interconnectedness.

The Earth Charter considers human rights as the basis of, and at the same time, the limitation to, human welfare and existence. It is based on the unity of human and non-human life. To this end, procedural and certain substantial human rights are strengthened, while other substantial human rights are limited. This is a novelty in international human rights law.

Some extracts can illustrate this (emphasis added):

Preamble

... We must join together to bring forth a sustainable global society founded on respect for nature, universal human rights, economic justice, and a culture of peace.

... The spirit of human solidarity and kinship with all life is strengthened when we live with reverence for the mystery of being, gratitude for the gift of life, and humility regarding the human place in nature.

... We affirm the following interdependent principles for a sustainable way of life as a common standard.

The actual Charter with its 16 principles contains references to both the strengthening aspects and the limiting aspects of human rights.

1. Strengthening of human rights in the Earth Charter:

Principle 3 (a)

Ensure that communities at all levels guarantee human rights and fundamental freedoms and provide everyone an opportunity to realize his or her full potential.

---

133   For a proposal with respect to the Austrian constitution see Pernthaler, P. (1992), 'Reform der Bundesverfassung', in Pernthaler, P., Weber, K. and Wimmer, N. (eds), *Umweltpolitik durch Recht-Möglichkeiten und Grenzen* (Wien, Manz), 10; with respect to the Swiss constitution see Bundesamt für Umwelt, Wald und Landschaft (eds) (1995), *Die Würde der Kreatur* (Gutachten).

134   Earth Charter, adopted in June 2000 in The Hague; see <www.earthcharter.org>, accessed 1 December 2007.

...

Principle 7

Adopt patterns of production, consumption, and reproduction that safeguard Earth's regenerative capacities, human rights, and community well-being.

Principle 8 (a)

Ensure that information of vital importance to human health and environmental protection, including genetic information, remains available in the public domain.

...

Principle 9 (a)

Guarantee the right to potable water, clean air, food security, uncontaminated soil, shelter, and safe sanitation ...

...

Principle 11

Affirm gender equality and equity as prerequisites to sustainable development and ensure universal access to education, health care, and economic opportunity.

Principle 12

Uphold the right of all, without discrimination, to a natural and social environment supportive of human dignity, bodily health, and spiritual well-being, with special attention to the rights of indigenous peoples and minorities.

Principle 13

 Strengthen democratic institutions at all levels, and provide transparency and accountability in governance, inclusive participation in decision making, and access to justice.

2. Duties with limitations to human rights in the Earth Charter:

Principle 1 (a)

Recognize that all beings are interdependent and every form of life has value regardless of its worth to human beings.

...

Principle 2 (a)

Accept that with the right to own, manage, and use natural resources comes the duty to prevent environmental harm and to protect the rights of people.

...

Principle 6 (a)

Place the burden of proof on those who argue that a proposed activity will not cause significant harm, and make the responsible parties liable for environmental harm.

Historically, the idea of human rights was shaped by two major political traditions, i.e. liberal and social thought. First, eighteenth century liberalism established the idea of individual freedom (French: *liberté*). Second, nineteenth and twentieth century democratic and social principles added the ideas of equality and solidarity (French: egalité and fraternité). To conceptualize human being as individuals in a free, democratic and social society has been the achievement of Modernity.

But time has moved on. While human beings continue to be a threat to themselves, they are increasingly threatening the natural conditions on which they depend. This calls for a broadening of the concept of solidarity. Future generations and the natural environment should be within the realm of solidarity.

Conclusion

The ecological approach to human rights acknowledges the interdependence of rights and duties. Human beings need to use natural resources, but they also completely depend on the natural environment. This makes self-restrictions essential, not only in practical terms, but also in normative terms. Entitlements to natural resources and a healthy environment, usefully expressed as rights, can no longer be perceived in purely anthropocentric terms.

Human rights, like all legal instruments, need to respect ecological boundaries. These boundaries can be expressed in ethical and legal terms as they define content and limitations of human rights. Will institutions be able to adapt to these new ecological human rights? In the interest of the law's coherence and efficiency, they ought to. In the interest of human survival, they must!

Chapter 5

The State as Environmental Trustee

Like the previous chapters on justice and human rights, the following chapter concerns a fundamental concept of law and governance. The state has the central authority to govern people in a given territory. The authority involves the making and enforcement of rules based on fundamental principles such as justice and human rights. Without the state, these principles could not be guaranteed. From the perspective of the principle of sustainability the question arises, therefore, how the modern state responds to the global environmental challenge and how this may affect its basic functions and duties.

This question has two very different dimensions. One concerns the state's internal (domestic) function, the other its external (international) function.

The domestic function of the state is determined by a constitution and performed through legislation. It is possible, therefore, to examine a national constitution and environmental legislation and see how issues of ecological sustainability are being served. As we have seen, most constitutions provide for environmental human rights or state obligations.

In formulating its constitutional goals, the state develops certain characteristics. These can be summarized by classifying the state as, for example, a 'constitutional state', 'social welfare state' or 'socio-constitutional state'. This way, a whole typology of state descriptions can be developed.[1] Emphasizing its environmental goals, the state has been classified as an 'environmental state' (German *Umweltstaat*[2]). Adding those attributes to the term 'state' does not in itself say much about the state's actual performance, but helps to identify conceptual approaches and differences. In the case of Germany, the term *Umweltstaat* has channelled and focused the debate on relevant principles, policies and laws. One key issue of debate here is the relationship between *Umweltstaat* and the *Rechtsstaat* ('constitutional state'). Are constitutional obligations to protect the environment in potential conflict with individual freedom and human rights, as some suggest,[3] or are those obligations an inherent part of the idea of freedom and human rights? This second approach is not liberal, but ecological in nature, leading to a different concept. Known in Germany as

---

1 Berg, W. (1997), 'Typologie von Staatsbeschreibungen' in Burmeister, J. (ed.), *Verfassungsstaatlichkeit. Festschrift für Klaus Stern* (Munich, C.F. Beck), 421, 425 et seq.

2 Kloepfer, M. (ed.) (1994), *Umweltstaat als Zukunft. Studien zum Umweltstaat* (Bonn, Economica); Calliess, C. (2001), *Rechtsstaat und Umweltstaat* (Tübingen, Mohr Soebeck Verlag).

3 Kloepfer 1994 (n. 2 above) and Calliess 2001 (n. 2 above).

ökologischer Rechtsstaat[4] or *ökologischer Verfassungsstaat*,[5] the 'eco-constitutional state'[6] or 'Green state'[7] describes the fundamental alternative to political liberalism with its creations of the modern *Rechtsstaat* and *Umweltstaat*. The ingredients are ecological sustainability, ecological justice and ecological human rights as described in the previous chapters. The difference between the two models is paradigmatic as we have seen, although commonalities and overlaps are possible in practical law-making.

The constitutional concept reflects the internal or domestic function of the state. However, we will not further discuss this here.[8] In this chapter we focus on the external or international function.

The international function is determined by international law. And as the state is defined in terms of international law with state sovereignty capturing both internal and external powers, its functions are mostly observed from an international point of view.[9] This point of view sets the 'state' in direct opposition to the global environment. The nation state represents fragmentation, the environment represents unity. They seem to come from totally different experiences. As Peter Sand observes: 'All revolutions have their iconoclastic phase. When international lawyers first embraced the global environmental revolution – looking for icons to smash – they were eager to pick on the nation state as a target'.[10]

Whether the environmental revolution may yet lead to a revolution of international law is an open question, but the state as such is not the target. For the most part, the debate has centred around territorial sovereignty. How can the state's territorial integrity be reconciled with the Earth's ecological integrity?

Fundamentally, there is a reality crisis. If the logic of ecological integrity is followed through, global governance based on sovereign nations appears unacceptable.[11] But without them no governance structures would be in place to develop the policies of ecological integrity. In search for the 'right' reality some compromise will have to be found, but a strong sense for changing realities is necessary as Joseph Camilleri and Jim Falk have noted:

---

4   Bosselmann, K. (1992), *Im Namen der Natur* (Munich, Scherz); Burmeister, H. (ed.) (1994), *Wege zum ökologischen Rechtsstaat* (Taunusstein, Eberhard Blottner).

5   Steinberg, R. (1998), *Der ökologische Verfassungsstaat* (Frankfurt, Suhrkamp).

6   Bosselmann, K. (1995), *When Two Worlds Collide* (Auckland, RSVP), 222–63.

7   Eckersley, R. (2004), *The Green State: Rethinking Democracy and Sovereignty* (Cambridge, MA, MIT).

8   For a discussion of one key aspect of the internal function of the state, i.e. the concept of citizenship, see the following chapter, Chapter 6.

9   From a citizen's perspective, their own state usually appears as 'the government'. From a legal perspective, a government represents the state, thus is different from it; the state, on the other hand, exists in a domestic as well as in an international context.

10   Sand, P. (2004), 'Sovereignty Bounded: Public Trusteeship for Common Pool Resources?', *Global Environmental Politics* 4, 47–71.

11   Lynton Caldwell, the great pioneer of environmental policy studies, has often remarked that political institutions are incapable of accepting ecological realities; see e.g. Caldwell, L.K. (1963), 'Environment: A New Focus for Public Policy?', *Public Administration Review* 23, 132–9.

The legitimacy of the sovereignty discourse resides in its capacity to explain reality, and as its ability to reflect upon the real situation of the world decreases, the erosion of its legitimacy increases. And, of all the areas which the sovereignty discourse ceased to be able to accurately clarify, it is precisely in the growing gap between the theorisation of reality it has attempted and the very reality of the ecological dynamics of the biosphere that this loss of legitimacy becomes clearer. [12]

Judging by states' behaviour, the sovereign discourse has not led us very far. Only too often states have followed the John Austin's definition of law, namely that law is the command of a sovereign backed by the threat or use of force.[13] In the current system of decentralized international law such legal positivism is hostile to the ideas of shared responsibility or future generations responsibility. For example, insisting on national interests or economic needs in climate change negotiations still reveals Austinian thinking, yet it is not uncommon and is, to an extent, accepted by all nations. Because states fear for their sovereign status, there is no sovereign in international law. Modern doctrines of international law, especially in the age of globalization, have moved beyond the idea that law only exists as a series of sovereign commands. But that is not the point. The problem is the ongoing gap between the theory of cooperation and the practice of competition.

A significant step towards closing this gap was the emergence of international environmental law since the 1970s. Initiated by the environmental movement, not by states, this new branch of international law emerged from changed realities. As the threats of the global environmental crisis were commonly felt, states found it necessary to cooperate. The rules, processes and mechanisms adopted by states since Stockholm and Rio are mostly those of cooperation, but they have not modified territorial sovereignty, nor were they designed to do so.

For good measure, the new principles promote cooperation, negotiation and consultation but, as we shall see, within a decentralized system of sovereign actors. There is no higher sovereign or principle that they are expected to follow.

The foundational norm of international environmental law is expressed in Principle 21 of the Stockholm Declaration:

> States have, in accordance with the Charter of the United Nations and the principles of international law, the sovereign right to exploit their own resources pursuant to their own environmental policies, and the responsibility to ensure that activities within their jurisdiction or control do not cause damage to the environment of other States or of areas beyond the limits of national jurisdiction.[14]

---

12 Camilleri, J. and Falk, J. (1992), *The End of Sovereignty?: The Politics of a Shrinking and Fragmenting World* (Aldershot, Edward Elgar), 171.

13 Austin, J. (1995), *The Province of Jurisprudence Determined*, ed. Rumble, W. (Cambridge, Cambridge University Press).

14 Declaration of the United Nations Conference on the Human Environment (Stockholm Declaration) (Stockholm, 16 June 1972; UN Doc. A/Conf.48/14, 11 ILM 1461 (1972)), Chapter I.

This text balances rights and responsibilities, but it does so on the basis of permanent sovereignty over natural resources.[15] This assumes a preference of exploitation over protection and a prerogative of national interests over global interests. The exact interpretation of Principle 21[16] is a matter of ongoing debate,[17] but it would be difficult to read a general prohibition of transboundary harm into this provision. Even if we assume its existence, the realization requires further negotiations and ultimately the application of the general rules of state responsibility.

These rules are guided by mutual respect of state sovereignty and the 'companionship of villains'[18] allowing each other a certain amount of environmental pollution. Liability only kicks in if clearly established thresholds are overstepped. Without treaties specifying such thresholds, states do not have to fear any consequences of transboundary environmental harm. The fact that no court in the world has ever convicted a state for breaching a duty to avoid environmental harm[19] speaks volumes. As Prue Taylor in her comprehensive analysis has shown, 'state responsibility for environmental harm is essentially a regime for the protection of property rights, rather than a regime for the protection of the environment *per se*'.[20] It undermines any effectiveness of a general duty not to harm and is 'ill-suited'[21] as a legal response to transboundary environmental pollution. This being the case, Principle 21 has not added anything to the well-established general rules of public international law.[22]

Other important principles of international environmental law include the principle of prevention, the precautionary principle, the principle of common concern of humankind and the principle of common but differentiated responsibility.

---

15   Schrijver, N. (1997), *Sovereignty Over Natural Resources: Balancing Rights and Duties* (Cambridge, Cambridge University Press); Chimni, B.S. (1998), 'The Principle of Permanent Sovereignty over Natural Resources: Toward a Radical Interpretation', *Indian Journal of International Law* 38, 208–17.

16   Principle 2 of the Rio Declaration is virtually identical (only referring to 'environmental and developmental policies'); Rio Declaration on Environment and Development (the Rio Declaration) (Rio de Janeiro, 13 June 1992; UN Doc. A/Conf.151/26 (vol. I); 31 ILM 874 (1992)).

17   Sands, P. (2003), *Principles of International Environmental Law*, 2nd edn (Cambridge, Cambridge University Press), 186; Ellis, J. and Wood, S. (2006), 'International Environmental Law', in Richardson, B. and Wood, D. (eds), *Environmental Law for Sustainability* (Oxford, Hart), 343, 357.

18   Bosselmann 1992 (n. 4 above), 125.

19   Effectively, the famous Trail Smelter Arbitration, *United States v. Canada* (1931–41; 3 Reports of International Arbitral Awards, 1938), still guides the rules of liability for transboundary environmental harm. Required is a wrongful act and an injury 'established by clear and convincing evidence'.

20   Taylor, P. (1998a), *An Ecological Approach to International Law* (London and New York, Routledge), 118.

21   Ibid. 123.

22   Ellis and Wood 2006 (n. 17 above), 358; See also Birnie, B and Boyle, A. (2002), *International Law and Environment*, 2nd edn (Oxford, Oxford University Press), 111.

Each has its own history and a more or less defined meaning,[23] and they all aim to restrict arbitrariness and carelessness of states. But none of these or other environmental principles[24] have ever amounted to a general duty not to harm. The concept of territorial sovereignty fundamentally stands against this. It guarantees the permanent sovereignty over natural resources and weakens the prospect for states taking responsibility for the global environment.[25]

Of course, environmental responsibilities can be negotiated and have, in fact, been negotiated in numerous environmental agreements. There are legally binding responsibilities with respect to specific areas such as the global climate, biodiversity, marine environment, and so on. But all these agreements have validated the sovereignty of states and their freedom to accept responsibility at their own discretion. States negotiate with others, but do not restrict themselves.

From a sustainability perspective, a duty to protect the environment should not have to be negotiated. Like justice and human rights, sustainability should principally be non-negotiable. As foundational norms of humanity, they need to guide the functions of states rather than be guided by them.

First Steps towards Redefining Territorial Sovereignty

In the following section we will develop the argument for a redefinition of territorial sovereignty. Essentially, the argument is that national territories are part of the global environment and therefore restricted in their use and exploitation. The effect of a redefined concept of territorial sovereignty would be a guardianship or trusteeship role for the state with respect to the (global) environment.

So how could the tensions between territorial sovereignty and the global environment be overcome? Fundamentally, we need to consider two factors that determine any changes in this relationship.

Firstly, there is a conceptual problem.[26] The indivisibility of the global environment is in stark contrast with the fragmentation of environmental law. While environmental thinking centres around the global environment, legal thinking centres around states. States create domestic law and international law alike. Domestically, environmental laws regulate certain segments of the environment (ignoring ecological integrity) and in competition with laws conducive to unsustainable development. Internationally, environmental laws represent the lowest common denominator among states and also allow the 'sovereign' state the choice to comply with them or not. Working together, these limiting aspects create a fragmentation and weakness of environmental law.

---

23   Kiss, A. and Shelton, D. (2004), *International Environmental Law*, 2nd edn (New York, Transnational Publishers), 247–94; Birnie and Boyle 2002 (n. 22 above), 112–52.

24   See above Chapter 2 and further below.

25   Bosselmann 1992 (n. 4 above), 122–36; Bosselmann 1995 (n. 6 above), 54–91; Taylor 1998a (n. 20 above), 110–43.

26   For the conceptual critique of environmental law see Bosselmann 1992 (n. 4 above), 108–75; Bosselmann 1995 (n. 6 above), 51–100.

Add to this the underpinning ethics of anthropocentric reductionism[27] and we are left with a global environment (in all its complexities) that stays almost unprotected.

The second factor is one of changing identity. The concept of state sovereignty is not static, but undergoing changes as states have to respond to ever changing circumstances. The 'changing concepts of sovereignty'[28] are widely noted in the literature. For example, Edith Brown Weiss's notion of 'operational sovereignty' as opposed to 'formal national sovereignty'[29] points to the fact that international environmental law has, in fact, restricted the exercise of sovereign powers. Even more restrictive has been the influence of economic globalization. Generally speaking, the globalization of the economy and ecology is the greatest challenge to the sovereign state since its inception through the 1648 Treaty of Westphalia.[30] In the age of globalization, clearly the modern territorial state is in search of a new identity.

The two factors are in conflict with each other. One is static, the other dynamic. We can also say that the first factor (the 'conceptual problem') is the legacy of the legal concept perpetuating its own methodology of law-making. The second factor ('changing identity') refers to forces determining the actual powers that the legal concept assumes. The question is, therefore, how the legal concept of state sovereignty should respond to changed realities. It can continue to ignore them, insisting on its own legitimacy and effectiveness. However, if it is true that national territoriality is at odds with the indivisibility of the global environment, the only option left is to readjust the legal fiction to ecological reality.

The point made so far is one of principle. There are no reasons in principle why the definition of territorial sovereignty should stay unchanged. As the concept of the sovereign state is dynamic, it is also open to new functions and responsibilities, as long as its basic functions to govern and serve common interests are not at risk.

We can now ask what these common interests are and what governance for sustainability entails.

To a certain extent, the development of international environmental law has shown that the legal fragmentation is neither complete nor absolute, but always relative to common interests (as perceived by states). Some form of limitation to territorial sovereignty for the sake of pursuing common interests is clearly accepted today.[31] And even though the concept of state sovereignty remains the great divide between national and international law, it is also true that the difference between these has become blurred.[32] It is at least conceivable that common interests necessitate rearrangements of sovereignty concepts.

---

27   Bosselmann 1992 (n. 4 above), 250–63; Bosselmann 1995 (n. 6 above), 40–50.

28   Kiss and Shelton 2004 (n. 23 above), 25.

29   Brown Weiss, E. (1993), 'International Environmental Law: Contemporary Issues and the Emergence of a New World Order', *Georgetown Law Journal* 81, 675–710.

30   The treaty text can be viewed at <http://www.yale.edu/lawweb/avalon/westphal. htm>, accessed 1 December 2007.

31   Kiss and Shelton 2004 (n. 23 above), 26; also further below.

32   Taylor, P. (2002), 'The Global Perspective: Convergence of International and Municipal Law', in Bosselmann, K. and Grinlinton, D. (eds), *Environmental Law for a Sustainable Society* (Auckland, New Zealand Centre for Environmental law), vol.1, 123–43.

The Importance of Common Interests

From a sustainability perspective, common interests suggest limitations to state sovereignty allowing for some form of global governance. From a perspective of states, common interests suggest less than that, perhaps some compromise between economic and environmental interests. The basis of state sovereignty as the divider between domestic and international environmental law would stay untouched. Either way, both perspectives claim to pursue 'comment interests' which leaves us with further questions on what 'common interests' ultimately are and how they ought to be identified.

'Common interests' are not the same as 'common interest'. The former is a sociological, the latter, a legal concept, but both are related. Common interests gain legal significance in so far they are recognized as relevant to legal norms. International environmental law operates with a whole host of terms signalling forms of a 'commonality of interests':[33] from mere common 'concerns',[34] 'global consensus'[35] or 'interest of the world'[36] to 'common interest',[37] 'common interest of mankind',[38] 'interest of all mankind'[39] or 'interest of mankind as a whole'[40] to the recent notions of 'common concern of mankind',[41] 'common concern of humanity'[42] and the concept of 'common heritage of mankind'.[43] Some of these expressions reflect different degrees of a commonality of interests;[44] however, they all presume the existence of a common interest in a legal sense.

---

33   Brown Weiss 1993 (n. 29 above), 710.

34   For example, 'concerns for sustainable development', Principle 1, Rio Declaration 1992 (n. 16 above).

35   Agenda 21: Programme of Action for Sustainable Development (Agenda 21) (Rio de Janeiro, 14 June 1992; UN Doc. A/Conf.151/26 (1992) 31 ILM 874 (1992)), Preamble.

36   For example, 'interest of the world in safeguarding for future generations', Convention for the Regulation of Whaling 1946, Preamble.

37   Ibid.

38   Tokyo International Convention for the High Seas Fisheries of the Northern Pacific Ocean 1952, Preamble.

39   Antarctic Treaty 1959, Preamble.

40   Madrid Protocol on Environmental Protection to the Antarctic Treaty 1992, Preamble.

41   United Nations Framework Climate Change Convention 1992 (New York, 9 May 1992; UN Doc. A/AC.237/18, 31 ILM 848), Preamble; Convention on Biological Diversity 1992 (Rio de Janeiro, 5 June 1992; 31 ILM 818 (1992)), Preamble.

42   IUCN Draft International Covenant on Environment and Development 1995, Article 3.

43   Moon Treaty 1957; UN Convention on the Law of the Sea 1982.

44   For example, Kiss considers 'common interest' to be stronger than 'common concern', but sees 'common concern of humanity' as a 'principle' or 'general concept', thus more compelling than both; Kiss, A. (1999), 'International Trade and the Common Concern of Humankind', in Bosselmann, K. and Richardson, B. (eds), *Environmental Justice and Market Mechanisms: Key Challenges for Environmental Law and Policy* (London, Kluwer Law International), 143 at footnote 3, 144 at footnote 2, 148–9, respectively; See also Brunee

If 'common interest' is a generic legal term for the various expressions of legal significance,[45] it can be identified as one of the key developments not only in international environmental law, but international law in general. Its history began after World War II when general obligations in the common interest of all humanity expanded in the area of human rights. The common interest found its strongest expression in global and regional human rights conventions and the recent development towards an 'extraterritoriality'[46] of human rights. The common interest has clearly promoted human rights as an essential need; it has also promoted environmental protection as an essential need, albeit to a lesser extent. There may be political and economic reasons for such a difference. Human welfare – as expressed in human rights – may be more consistent with Western values than ecological welfare. Nevertheless, both reflect values, and we may ask whether the common interest is deemed to continue the humanist tradition of excluding the ecological context of human welfare. If it is true that the natural environment is the basis for all life and the ecology the basis of all economy, it seems not unreasonable to value the environment in similarly emphatic terms as human rights or state sovereignty. Common interest may be able to create acceptance for the idea of environmental governance.

Common Interest in Governance for Sustainability?

The proposed argument for a new form of governance is as follows. The development of international environmental law is characterized by the recognition of a common interest in protecting the global environment. This recognition, visible in new principles and concepts, has effectively limited state sovereignty. It has not, however, principally limited or questioned territorial sovereignty. The failure to do so perpetuates the gap between legal fiction and ecological reality. In order to close this gap, territorial sovereignty needs to be redefined by exempting transnational aspects of the domestic environment from the concept of territorial sovereignty.

Territoriality, in its classic form, is outdated.[47] It is no longer the state's exclusive domain and not the defining moment for the state's identity. Borders have not only become permeable to human, material and intellectual exchanges, they have increasingly lost their function to secure territoriality. By their very nature modern weaponry, terrorism, communication technology, free trade, the environment and human rights ignore national boundaries. Exercising territorial sovereignty does little to protect the state's enclosure against undesirable invasions.

J.(1989), 'Common Interest – Echoes from the Empty Shell?', *ZaöRV* 49:4, 791–808, 807; and Taylor 1998a (n. 20 above), 278.

45   Following Kiss, A. (1985), 'The Common Heritage of Mankind: Utopia or Reality?', *International Journal* 40, 423–41, 428; and Taylor 1998a (n. 20 above), 273, 277.

46   Meron, T. (1995), 'Extraterritoriality of Human Rights Treaties', *American Journal of International Law* 89, 78–82. While the March 1999 NATO intervention in Kosovo may have been in violation of the UN Charter and the national territoriality of the Federal Republic of Yugoslavia, it may also have indicated a new stage of human rights limitations to state sovereignty.

47   Bosselmann 1995 (n. 6 above), 74–94.

Nowhere is the gap between legal fiction and reality more striking than with respect to the environment. What a state does, in and to its environment, affects not only its own territory, but the territory of other states together with the planetary environment shared by all. The 'export' of environmental interferences into the 'sovereign' territory of other states is matched by the 'import' of other states' environmental interferences into their own 'sovereign' territory. Such practice of invasion and retaliation may be acceptable by the international community as a matter of inevitability. This raises, however, a serious concern. The mere exercise of permanent sovereignty of states over their natural resources leads to the 'tragedy of the commons' (Hardin 1968). The mutual acceptance of absolute territorial sovereignty and the environmental degradation associated with it undermines the common interest in protecting the global environment. Would it not be better, then, to define territorial sovereignty in a less absolute manner?

Mere contracting between states will not do. The present system of treaties, customary rules and general principles has produced many norms and obligations: reciprocal or regulatory, *inter partes* or *erga omnes*, legal or moral. But none of them has ever challenged the state's discretionary power to exercise its absolute and complete sovereignty over its own piece of territory. Under the regiment of territorial sovereignty the protection of the global environment is deemed to stay fragmented and departmentalized.

Closing the gap between legal fiction and reality requires, therefore, adopting an idea that better matches reality: ecological sustainability and territorial sovereignty must not compete, but must complement each other. To advance this idea, it will be helpful to recall some absurdities of the existing regime.

Environmental Challenges to Territorial Sovereignty

Territorial Sovereignty over the Environment

Any community, including the community of nations, is shaped by certain values or beliefs. Modern international law is based on the central belief that states represent the interests of their citizens and that international relations are best served by a legal system based on the respect for equality of sovereign members (Article 2(1) UN Charter) and on the principle of non-intervention in internal affairs (Article 2(7) UN Charter).

Historically, territorial sovereignty emerged as a means to protect state boundaries, not the use of territory or resources. When Hugo Grotius (1583–1645) formulated the imperative of international law 'that foreign property is respected' (*De jure bellis ac pacis*), he saw the exclusive competence of the state over its territory essentially as the protection of property. With respect to nineteenth century international law theory, Anthony Carty found this link firmly established: 'The relationship of a state to its own territory could be described in the same language as that of a private individual towards his own property, i.e. in the sense of a spatial dimension over which he had absolute right of use and disposal'.[48] The language may be different in international documents, but economic interests were clearly behind the reasoning for the mid twentieth century concepts of the 'right to exploit freely the natural resources'[49] and the permanent sovereignty over natural resources.[50] Around the time of the 1972 Stockholm Conference, additional documents[51] broadened the concept to specifically include all economic activities of states and any property rights associated with them.

Property rights thinking influenced the drafting of Principle 21 of the Stockholm Declaration ('States have ... the sovereign right to exploit their own resources')[52] and formed the background against which Principle 21 added a duty to protect the environment, as mentioned earlier. The International Court of Justice found this duty to be 'now part of the corpus of international law relating to the environment'[53] emphasizing the 'great significance that it attaches to respect for the environment, not only for states but for the whole of mankind'.[54]

Despite such broad respect for the environment, the actual obligations are much more narrowly defined, as discussed. They are hampered by rigid territorial sovereignty on the one hand and the continued legal insignificance of the environment on the other. Let us now examine this in more detail.

Legally the 'environment' only exists as areas within or outside national jurisdictions. Thus, not the environment, but state-owned parts of it are the referential point for any environmental obligation. Viewing the environment solely from the perspectives of states has an absurd effect: it creates the perception of four different environments. One environment is owned by the state, a second environment owned by other states, a third environment is owned by all (high seas, Antarctica, superadjacent airspace) and a fourth, global environment is owned by no one. Consequently, the first environment is of concrete, immediate interest, the second environment of still concrete, but less immediate interest, the third of a more abstract interest and the fourth environment of an even more abstract, distant interest.

---

48   Carty, A. (1986), *The Decay of International Law?* (Manchester, Manchester University Press), 44; See also Taylor 1998 (n. 20 above), 119–20.

49   First formulated in UN General Assembly Resolution 523 (VI) of 12 January 1952, United Nations Yearbook 5 (1951), 418; and UN General Assembly Resolution 626 (VII) of 21 December 1952, United Nations Yearbook 6 (1952), 390.

50   Beginning with the UN General Assembly Resolution 837 (IX) of 14 December 1954, United Nations Yearbook 12 (1958), 212, nine further UN resolutions between 1958 and 1974 established this concept as a heritage of newly emerging international environmental law.

51   For example, the Charter of Economic Rights and Duties of States (Article 2 II), UN General Assembly Resolution 3281 (XXIX) of 12 December 1974 (International Legal Materials 14 (1975)), 251.

52   Taylor 1998a (n. 20 above), 119–20. The desire of the new developing states for greater economic independence was, of course, the other driving force behind the concept of permanent sovereignty over natural resources.

53   Legality of the Threat or Use of Nuclear Weapons, ICJ Advisory Opinion (8 July 1996), 241–2, para. 29.

54   *Case Concerning the Gabçikovo-Nagmaros Project* (*Hungary/Slovakia*), 1997 ICJ, 37 ILM 162 (1998), para. 53.

This hierarchy of interests is matched by a hierarchy of obligations. The first environment is governed by sovereign rights (to exploit, use or preserve), the second is governed by an obligation to 'not cause damage', the third by a more abstract duty not to harm and the fourth is governed by a mere obligation to cooperate.

The nature and significance of environmental obligations depends entirely on which environment is affected by activities.[55] There is no uniform definition for activities harmful to the environment. Most common is the notion of 'pollution' as, for example, defined by the OECD Council Recommendation of 1974:

> Pollution means the introduction by man, directly or indirectly, of substances or energy into the environment resulting in deleterious effects of such nature as to endanger human health, harm living resources and ecosystems, and impair or interfere with amenities and other legitimate uses of the environment.[56]

The various elements of this definition[57] appear in many international texts according to their specific focus.[58] They cover transboundary and long-range pollution, but also pollution limited to the state's jurisdiction. The definition is fairly broad, and thus may be useful as a benchmark for assessing the various obligations following the four kinds of environment being affected by pollution. We can distinguish between intra-territorial, transboundary, common areas and global forms of pollution.

(1) Intra-territorial pollution   Intra-territorial pollution is confined to the state's territory[59] where it originates. Essentially, the control of this form of pollution is an internal affair. International environmental law knows no obligation not to pollute the 'national' environment, nor is there an obligation to protect it.[60]

The only way for a state to accept environmental obligations for its territory has been via specific treaty law. Some conventions for the protection of living organisms and the natural heritage contain obligations to preserve national resources. The Convention on Biological Diversity, for example, imposes the general obligation on

---

55   Since the 1960s international environmental law has created numerous environmental obligations which effectively limit the exercise of sovereign rights over the environment. In addition to older limitations related to transboundary pollution ('principle of good neighbourliness', '*sic utere tuo ut alienam non laedas*') new concepts such as the cooperation principle, the polluter pays principle, the precautionary principle, various forms of shared resource use and duties to protect the environment have been introduced. Together they make up 'the international common law of the environment'; Kiss and Shelton 2004 (n. 23 above), 247–94.

56   OECD Council Recommendation C(74)224 of 14 November 1974.

57   For an analysis see Kiss and Shelton 2004 (n. 23 above), 268–70.

58   With respect to the marine environment e.g. UN Convention on the Law of the Sea 1982, Article 1(1)(4).

59   Including territorial land and territorial seas together with the airspace (of about 90 km) above.

60   Kiss, A. (1989), 'Nouvelles tendances en droit international de l'environment', *German Yearbook of International Law* 32, 241–63, 258; Wolfrum, R. (1990), 'Purposes and Principles of International Environmental Law', *German Yearbook of International Law*, 308–30, 328.

states to take effective national action to halt the destruction of species, habitats and ecosystems. States parties are to apply Convention requirements inside the territorial limits even if there are no (obvious) transboundary or global effects.[61] Similar requirements exist under the Bern Convention on the Conservation of European Wildlife and Natural Habitats. States parties have an obligation to conserve wild flora and fauna in all circumstances, whether the problems posed are transboundary or not.[62] Finally, the UNESCO Convention for the Protection of the World Cultural and Natural Heritage[63] is based on the recognition that some cultural or natural heritage of various nations may have value as world heritage of humankind as a whole.[64] Elements of the world natural heritage that are situated on a state party's territory are, however, still governed by territorial sovereignty and property rights.[65] The only limitations to sovereignty are obligations to ensure conservation for future generations, to take appropriate measures and to report periodically.

In some way the UNESCO Convention is unique as it controls intra-territorial cultural and natural heritage that may not have any ecological connection with the rest of world. However, the Convention clearly depends on the state's discretion to identify suitable sites and fully respects territorial sovereignty. Even if one stresses the fact that treaty law has imposed environmental obligations on intra-territorial areas or pollution, it would be very difficult to derive a customary rule or general principle from that. There are no indications in the practice of states or the treaty-making process to suggest such a general obligation.

Among customary rules and general principles relevant in respect to intra-territorial pollution are the general duty to cooperate, the good faith principle and perhaps the precautionary principle. However, being abstract they provide little content for environmental obligations, and none of these principles would have the effect of limiting territorial sovereignty.

(2) Transboundary impacts   The term of 'transfrontier'[66] or 'transboundary'[67] commonly refers to pollution that originates (wholly or in part) within the area under

---

61   It should be noted that negotiators rejected proposals to consider biological diversity as a common heritage because of concern for national sovereignty.

62   The Convention on the Conservation of European Wildlife and NaturalHabitats (Bern Convention) (Bern, 19 September 1979; European Treaty Series 104), Article 2.

63   UNESCO Convention for the Protection of the World Cultural and Natural Heritage (Paris, 23 November 1972; 11 ILM 251).

64   Kiss and Shelton 2004 (n. 23 above), 331 at footnote 87, note that 'significantly' the Convention maintains the approach that the cultural and natural heritage constitutes 'property'.

65   Convention for the Protection of the World Cultural and Natural Heritage (n. 63 above), Article 6(1).

66   For example, Implementation of a Regime of Equal Right of Access and Non-Discrimination in Relation to Transfrontier Pollution, OECD (Recommendation C (77) 28 (1977)).

67   For example, the Convention on Long-Range Transboundary Air Pollution (Geneva, 13 November 1979; 18 ILM 1449).

the jurisdiction of one country and which has effects in the area under the national jurisdiction of another country. Essentially, the presence of two or more countries is required here. Transboundary effects may concern air, water, soil or biodiversity and there are many treaties covering its various forms of transport and effect.[68] Likewise, several customary rules and general principles with limiting effects on the exercise of territorial sovereignty are applicable.

A number of such limitations follow from the fact that the 'receiving' state has the same range of exclusive territorial competences as does the polluting state. For example, the doctrine of abuse of rights (*sic utere iure tuo ut alterum no laedas*) prohibits the arbitrary exercise of sovereignty as it would undermine the ability of other states to fully exercise their own sovereignty. Consequently, the OECD Principles on Transboundary Pollution demand 'a fair balance of the rights and obligations among countries concerned by transfrontier pollution'.[69]

A prime example of balancing competing sovereign rights are the principles governing neighbouring states. The right to remain free from outside intervention, in particular with respect to transboundary pollution, is firmly established since the Trail Smelter Arbitration.[70] This arbitration is often considered the *locus classicus* of international environmental law[71] as it was instrumental for the establishment of state responsibility for environmental harm and also for states' obligation to prevent activities within their jurisdiction or control.[72] Further classic cases such as Corfu Channel[73] and Lac Lanoux[74] contributed to the recognition of the right not to suffer damage from transboundary pollution.

It is important to note though that all these obligations do not aim to protect the environment *per se*, but to protect the territorial integrity of neighbouring states. Any limiting effects to the territorial sovereignty of the polluting state are solely due to the territorial sovereignty of the 'receiving' state.[75]

---

68   See for example Kiss and Shelton 2004 (n. 23 above), 266–294, 395–432, 527–58. A notable exception to an otherwise broad coverage is the protection of soils; apart from the 1994 UN Convention to Combat Desertification, UNEP guidelines for national soil policies and an FAO Code of Conduct on the Distribution and Use of Pesticides there are no agreements controlling transboundary effects on soils.

69   Principles Concerning Transfrontier Pollution, OECD (Recommendation C(74) 224, 14 November 1974), Introduction.

70   (1941) *American Journal of International Law* 35, 684–734.

71   For example Handl, G. (1975), 'Territorial Sovereignty and the Problem of Transnational Sovereignty', *American Journal of International Law* 69, 50–76, 60.

72   Taylor 1998a (n. 20 above), 65–9; Kiss and Shelton 2004 (n. 23 above), 274–7.

73   Adding the state's obligation not to allow knowingly its territory to be used for acts contrary to the rights of other states; the Corfu Channel Case (*United Kingdom of Great Britain and Northern Ireland/Albania*) ICJ Reports 1949, 4.

74   Alluding to the problem of pollution of shared waters; (1959) *American Journal of International Law* 53, 156.

75   Odendahl, K. (1998), *Die Umweltpflichtigkeit der Souveränität* (Berlin, Duncker & Humblot), 158; Bosselmann, K. (2004), 'Environmental Governance: A New Approach to Territorial Sovereignty', in Goldstein, R.J. (ed.), *Environmental Ethics and Law* (Aldershot, Ashgate), 293–313, 124; Handl 1975 (n. 71 above), 52.

The same logic applies to the concept of shared natural resources. The 1973 UN Resolution 'Co-operation in the Field of the Environment concerning Natural Resources shared by Two or More States'[76] shifted the focus from mere pollution control to the protection of natural resources. However, the underlying justification was and is the concern for territorial integrity. As mentioned above,[77] the duty of states to ensure that activities within their jurisdiction or control do not cause damage to the environment of other states (Principle 21 of the Stockholm Declaration) is, essentially, a duty to respect territorial sovereignty and property rights. The idea of actually limiting these rights in order to accommodate genuine environmental protection is alien to positive international law.

Other principles like good neighbourliness, good faith, cooperation, solidarity, polluter pays or precautionary all have their importance for the relationship between states, but they do not create any limitations to territorial sovereignty.[78] All obligations in relation to transboundary pollution are concerned with a fair balance of use and exploitation rights rather than the environment. Environmental protection remains a mere side effect.

(3) Common areas impacts   Transboundary pollution affecting not other states, but areas under no state jurisdiction, can be categorized as common areas pollution. The areas concerned here are the high seas (including the seabed), Antarctica and the superadjacent airspace (not including outer space[79]). They are distinguished from the global environment not only by their spatial nature, but also by the fact that international law recognizes them in a specific manner.

The 'common areas'[80] are recognized either as humankind's common heritage (seabed), common concern (Antarctica) or shared resources (high seas, airspace). The common interest behind each of these concepts varies in weight and so do the corresponding obligations: the environments of the seabed and Antarctica are, in theory, better protected, imposing more limitations to sovereignty and property rights than the high seas and the airspace.

The common areas are referred to in Principle 21 of the Stockholm Declaration as 'areas beyond the limits of national jurisdiction'. Principle 21 protects the common areas only against effects coming from activities within a state's jurisdiction. However, obligations also exist in respect of activities within the common areas. Articles 192–5 of the UN Convention on the Law of the Sea, for example, formulate the duty

---

76   UN General Assembly Resolution 3129 (XXVIII) of 13 December 1973.

77   [*Territorial Sovereignty over the Environment* above, 154].

78   Odendahl 1998 (n. 75 above), 205–30; Taylor 1998a (n. 20 above), 123.

79   The exclusion of outer space (beyond an altitude of 90–100 km) is justified because there is no measurable evidence of pollution originating from countries. Debris and chemical emissions originate from activities within space; other chemical or radioactive pollution has not been found.

80   This term is preferable to the term 'global commons' which is sometimes understood as a plural term for the various areas and aspects beyond national jurisdiction (including atmosphere or global biodiversity), sometimes as a singular term for all these areas and sometimes in the sense of 'common areas' as defined here. For an analysis see Taylor 1998a (n. 20 above), 165–9.

of states to protect and preserve the marine environment regardless of the locus of activities. In the case of the seabed (including ocean floor, subsoil and resources) Article 136 declares the 'area' as the *common heritage of mankind*. The meaning and significance of the common heritage of humankind is, of course, far from certain, but the common interest behind it clearly acknowledges the interdependence of the Earth's ecosystems, the interconnectedness of activities of all humanity and the need for comprehensive management.[81] The associated level of environmental protection is, therefore, considerably higher than that for the high seas or the airspace, for example.

Outside the binding effects of treaty law the concept of common heritage has no legal status; it is neither a customary rule nor a general principle. This does not render it toothless. The common heritage concept provides an important basis for governance deriving from legal principles of property and public trust.[82] However, the concept cannot in any way limit territorial sovereignty mainly because it only concerns activities within common areas not within the states' territory. The situation is similar in respect of the protection of Antarctica. The Antarctic Treaty system, even with its most recent addition to the 1959 Antarctic Treaty[83] and the 1980 Canberra Convention,[84] namely the 1991 Madrid Protocol on Environmental Protection,[85] is far from certain. On the one hand, the 1991 Protocol emphasizes the 'intrinsic value of Antarctica' and 'interest of mankind as a whole' achieving full recognition of the common interest.[86] On the other hand, the treaty system does not guarantee protection forever; there are too many loopholes and opting-out possibilities. Given such uncertainty a customary rule with limiting effect to the states' territorial sovereignty could not have emerged.

With respect to the marine environment, the prohibition on polluting the high seas is considered a general principle of international law.[87] The problem here is its generality. To result in liabilities for states or limitations to their sovereignty, the principle requires specific treaty regulations.

Finally, the importance of Principle 21 for the protection of common areas should be noted. The duty not to harm the environment of common areas goes further than the well-established customary rules of transboundary pollution. However, it is doubtful that this new duty is a rule of customary international law.[88] The mere repetition of such a duty in recent treaties is in itself not enough, and even if we

---

81   Taylor 1998a (n. 20 above), 278.

82   Kiss and Shelton 2004 (n. 23 above), 248.

83   Antarctic Treaty (Washington, 1 December 1959; 402 United Nations Treaty Series, 71).

84   Convention on the Conservation of Antarctic Marine Living Resources (Canberra Convention) (Canberra, 20 May 1980; 19 ILM 841).

85   Protocol on Environmental Protection to the Antarctic Treaty (Madrid Protocol) (Madrid, 4 October 1991; 30 ILM 1461).

86   Kiss and Shelton 2004 (n. 23 above), 252; Kiss, A. 1999 (n. 44 above),147.

87   Kiss and Shelton 2004 (n. 23 above), 261.

88   Odendahl 1998 (n. 75 above), 231–2, 260–61.

assume a customary rule[89] or a 'general obligation',[90] it would be limited to those common areas covered by treaty law.

The overall picture shows few signs of a limitation to territorial sovereignty with respect to common areas. With the exception of an obligation to protect the marine environment,[91] no general environmental obligations exist. The reason for this is that the areas beyond national jurisdictions are not categorized in total, but individually, with differing obligations for each of them.

(4) Global impacts   Global environmental pollution affects the environment which is neither confined to national jurisdictions nor to the areas beyond national jurisdictions. The global environment includes both, but embraces the earth as a whole. Any limitations to sovereignty to be considered here would not originate from territorial concerns of states, but from the global environment *per se*.

The World Charter for Nature of 1982 was the first document to focus on the earth as a whole. It aims to protect the global environment for its own sake, independently from jurisdictions or spatial segments. The UN resolution carrying the Charter[92] was opposed by the United States, and its principles were not developed further in a binding legal instrument. But it helped considerably to give international environmental law direction and shape.[93] With its emphasis on the intrinsic value of nature and the need for humanity to be guided by a code of ethics, the Charter promoted ecocentrism as a viable alternative to anthropocentrism.[94] Ecocentrism, with its central notion of intrinsic values, increasingly influenced national[95] and international environmental law. Examples of international documents include the 1980 Convention for the Conservation of Antarctic Living Resources,[96] the 1991 Protocol to the Antarctic Treaty,[97] the 1979 Convention on the Conservation of European Wildlife and Natural Habitats,[98] the 1992 Convention on Biological Diversity and the 1995 IUCN Draft International Covenant on Environment and Development.[99] Article 3 of the Draft Covenant specifically defines the 'global environment' as a 'common concern of humanity'.

---

89   Kiss and Shelton 2004 (n. 23 above), 281.

90   Legality of the Threat or Use of Nuclear Weapons (n. 53 above), 241–2, para. 29.

91   Birnie and Boyle 2002 (n. 22 above) speak of the 'emergence' of such an obligation (351) and note that it 'reflects its extension to global common areas contemplated by Principle 21 of the Stockholm Declaration' (352), but also point out 'unhelpful generalities' (353) and the generally restricted concept of liability (382–91).

92   UN General Assembly Resolution 37/7 of 28 October 1982.

93   Kiss and Shelton 2004 (n. 23 above), 65.

94   Taylor 1998a (n. 20 above), 300.

95   For example the New Zealand Resource Management Act 1991, sections 2 and 7; see above, Chapters 3 and 4.

96   Canberra Convention (n. 84 above).

97   Madrid Protocol (n. 85 above).

98   Bern Convention (n. 62 above).

99   See Bosselmann, K. (2006), 'Ecological Justice and Law', in Richardson, B. and Wood, S. (eds), *Environmental Law for Sustainability: A Critical Reader* (Oxford, Hart Publishers), 129, 161–2.

Another significant move towards a fundamental obligation to protect the global environment has been the development of the Earth Charter. Representing a consensus created by global civil society and endorsed by UNESCO and IUCN, the Charter's fundamental principles of respect and care for the community of life and ecological integrity are the most powerful expression of global responsibility to date.

The World Charter for the Conservation of Nature, the Earth Charter and the IUCN Draft Covenant are milestones for a future global environmental treaty and/ or the recognition of sustainability as a fundamental principle. However, due to their lacking normative character, they have currently no direct effect on territorial sovereignty.

Such an effect could be generated by treaties covering the global environment. At present, however, treaties only exist in respect of some of the most urgent threats to the global environment: loss of biodiversity, ozone layer depletion and climate change. The 1992 Biodiversity Convention, the 1985 Convention for the Protection of the Ozone Layer and the 1992 Framework Climate Change Convention – each with their own regime of ongoing negotiations – cover important aspects of the global environment. But none of these regimes intend any restrictions to the sovereignty of states over their national environment. They rely on cooperation, negotiation and good faith, using traditional means of compliance.

Conclusion

International environmental law has created some important new obligations beyond treaty law. The various forms of these obligations set certain limits to the state's sovereignty. Yet the current system of environmental limitations to territorial sovereignty reveals major deficits.[100]

First, environmental limitations do not protect the environment *per se*, but territorial interests in the environment. Direct protection of the environment is the aim of strategies to protect the global environment; attempts to this effect have, however, not reached hard law status. Second, the degree of environmental limitations depends on the space affected by pollution (state jurisdictions, common areas or global environment). Such variation is not justified considering that the environment is affected as a whole and not in artificially defined segments. Third, territorial sovereignty is not even restricted for the gravest forms of pollution, the pollution of the global environment; existing treaty law imposes no limitation on states' discretion.

The consequences of the current state-centred, anthropocentric system for the quality of environmental protection are serious and far-reaching. The focus on territorial sovereignty leads to a domination of rich states over poor, of today's interests over tomorrow's and of human needs over environmental needs. This 'logic of self-extermination'[101] is bound to fail and must be replaced by a different logic.

---

100   See also Taylor 1998a (n. 20 above), 123–4, 182–3; Odendahl 1998 (n. 75 above), 311–13.
101   Bosselmann 1995 (n. 6 above), 40.

Reconciling Territorial Sovereignty and Ecological Sustainability

In some way, governance for sustainability and state governance mark opposite poles. Either the environment or the state is the starting point for defining sustainability strategies. The first option is, of course, purely hypothetical if we think of the environment as the natural (i.e. non-human) environment. However, if we liken the 'environment' to the Earth's integrity, we may be better able to capture a unity of human beings, states and the natural environment and more readily accept the idea of sustainability governance. In this sense, governance for sustainability is not opposed to traditional forms of governance, but offers a broader outlook with additional aspects to consider. From a sustainability viewpoint, effective governance calls for an integration of the global environment into the concept of territorial sovereignty. The externalized environment becomes internalized.

The internalization or integration of the environment requires certain changes to the definition of territoriality. Sovereignty over the own 'national' environment is to be limited by the fact that the 'national' environment is also part of a wider, transnational or global environment.

Several recent trends in international law can help to formulate a more comprehensive limitation of territorial sovereignty. They may not all follow the same intention, but seen together they indicate a strong desire to break the monopoly of the sovereign state in defining scope and extent of environmental strategies. However, an important distinction has to be made between trends emphasizing the common interests of states and other trends emphasizing common interests in the (global) environment. The former approach seeks an expansion of existing instruments and principles on the basis of territorial sovereignty, the latter amounts to environmental governance redefining territorial sovereignty. While both approaches do not necessarily oppose each other, it will be seen that only a sustainability governance approach would overcome the systemic deficits and failures discussed above.

The Reformist Approach

The International Law Commission has, for some time now, attempted to expand the concept of state responsibility (for wrongful acts) to include 'liability for injurious consequences arising out of acts not prohibited by international law'.[102] The 1989 ILC report contained a passage that set the scene for the search for environmental governance:

> actions within one or more countries may have profound effects upon the environment elsewhere, which is at the core of global change. This suggests that from an environmental perspective, the planet is a global commons, in which case there is no point to targeting the global commons for separate treatment. The recognition that the whole planet is environmentally a global commons may be exceedingly important, but the point needs to be made and argued explicitly and perhaps through the use of other nomenclature, such as common patrimony.[103]

---

102   ILC Report 1993 (UN Doc. A/CN 4/450).
103   ILC Report 1989 (UN Doc. A/CN 4/423), 22 (54).

Be it 'common patrimony' or 'environmental governance' or some other term that captures the state's responsibility for the global environment, the ILC's internal controversy on the merits of sustaining the traditional state sovereignty concept illustrates the dilemma.[104] There is no reconciliation between ecological reality and territorial sovereignty. The completed set of draft articles for state responsibility,[105] adopted in 1996, is a further illustration of this.

One obvious solution would be to establish the new governance form by way of a global treaty. There is certainly no shortage of pleas for a treaty solution of this nature,[106] and the existing 1995 IUCN Draft and 2000 Earth Charter are important models for future development.[107] It cannot be overlooked, though, that a treaty solution is a bit of a 'pact with the devil': how can states be expected to give up their sovereign rights which allow them full negotiations in the first place?[108] As Ulrich Beyerlin and Thilo Marauhn observe, 'the more substance the rules of an "umbrella treaty" have, the less initiative and creativity states would be able to show in the individual treaty shaping process'.[109] The main significance of treaty suggestions and models is of educative nature. They express a growing common concern and interest in environmental governance.

The two concepts of 'shared natural resources' and 'common heritage of humankind' could be expanded to cover the global environment as a whole. Conceptually, 'shared natural resources' may include shared fresh water, international watercourses,[110] underground water,[111] marine water,[112] soil,[113] air,[114] the atmosphere,[115] common areas or global commons,[116] or perhaps the entire global environment. And

---

104   For a detailed account see Taylor 1998a (n. 20 above), 144–83.

105   Draft Article on State Responsibility 1996 (UN Doc. A/CN 4/L.528/Add.2).

106   For example Theutenberg, B. (1984), 'The International Environmental Law – Some Basic Viewpoints', *Acad Droit Colloq*, 233; Brown Weiss, E. (1995), 'Environmental Equity: The Imperative for the Twenty-First Century', in Lang, W. (ed.), *Sustainable Development and International Law* (Dordrecht, Martinus Nijhoff), 26; Taylor 1998a (n. 20), 305–10.

107   Note the Earth Charter's central Principle 1 'Respect Earth and life in all its diversity'; similarly, the IUCN Draft's Article 2 'Respect for all life forms'.

108   Which is why the Earth Charter, negotiated by non-governmental organizations, is not designed as a draft covenant, but as a stand-alone 'peoples' charter'.

109   Beyerlin, U. and Marauhn, T. (1997*),* *Law-Making and Law-Enforcement in International Environmental Law after the 1992 Rio Conference* (Berlin, Erich Schmidt Verlag), 26.

110   Kiss and Shelton 2004 (n. 23 above), 399, 407, 582.

111   Barberis, J. (1988), 'Le régime juridique international des eaux souterainnes', *Annuaire Français de Droit International* 33, 129–62, 138, 153.

112   Kiss and Shelton 2004 (n. 23 above), 435.

113   Ibid. 387.

114   Handl, G. (1986), 'National Uses of Transboundary Air Resources', *Natural Resources Journal* 26, 405.

115   Riphagen, W. (1980), 'The International Concern for the Environment as Expressed in the Concepts of "Common Heritage" and of "Shared Natural Resources"', in Bothe, M. (ed.), *Trends in Environmental Law and Policy* (Berlin, Beiträge zur Umweltgestaltung), 342, 346; Odendahl 1998 (n. 75 above), 316.

116   Taylor 1998a (n. 20 above), 165–7.

equally, the 'common heritage' concept has been used to cover not just ocean floor and sea bed, but the high seas, Antarctica, the stratosphere, common areas[117] and also the climate,[118] the ozone layer, biological diversity[119] and forests.[120] Again, the entire global environment might be considered to be the common heritage of humanity.[121]

Such expansions could gradually lead to a fundamental duty to protect the global environment. Various authors have, therefore, favoured the idea of a 'shared environment' as a central reference for environmental obligations.[122] However, the underlying resource-oriented and anthropocentric approach remained unquestioned. To assume the preference of states' resource interests and not to recognize a common interest in the protection of the environment for its own sake, reveals more than an ethical difference. Considering the strong element of shared property underlying the 'common interest' expressed in both concepts,[123] the implication of instrumental, economic value will remain unchanged no matter how broad 'shared resources' or 'common heritage' are being perceived. To change their possessive nature, the 'common interest' should be interpreted as an interest in the environment for its own sake. A recognized intrinsic value of the environment would, for example, transform the concept of 'common heritage of humankind' to 'common heritage of all life'.[124] The common interest underlying this transformed principle rejects the anthropocentric utilitarian view that the planet exists purely for humans' sake.

As discussed in the previous chapter, the same fundamental ethical conflict is also visible with respect to human rights. The international protection of both human rights and the environment is clearly in the common interest, but also represents potentially conflicting values. To a degree, the conflict potential can be controlled through a human right to a healthy environment. As long as environmental rights are moulded in the tradition of individual protection against the state they make sense; they empower civil society to greater participation in environmental decision-making. However, the ethical *and* political conflict between rights *to* the environment and rights *of* the environment becomes evident once the effects of human rights on the degradation of the environment are taken note of. There is a clear causal link

---

117   Kiss, A. (1983), 'The International Protection of the Environment', in MacDonald, R. and Johnston, D. (eds), *The Structure and Process of International Law* (The Hague, Martinus Nijhoff), 1069, 1083.

118   Ibid.; see also agenda item 148 'Conservation of the climate as part of the common heritage of mankind', for the UN General Assembly (30 November 1988, UN Doc. A/43/905).

119   Kiss, A. (1982), 'La notion de patrimoine commun de l'humanité', *Recueil des Cours à l'Académie de Droit International* II, 99–256, 190, 193.

120   Stocker, W. (1993), *Das Prinzip des Common Heritage als Ausdruck des Staatengemeinschaftsinteresses im Völkerrecht* (Zürich, Schulthess Juristische Medien), 214.

121   See Taylor 1998a (n. 20 above), 290–91 with reference to discussions surrounding UN General Assembly item 148 (n. 118 above).

122   Handl 1975 (n. 71 above), 53; Kirgis, F.L. (1972), 'Technological Changes to the Shared Environment', *American Journal for International Law* 66, 290–320; Picht, G. (1971), 'Umweltschutz und Politik', *Zeitschrift für Rechtspolitik* 4, 152–8.

123   [See above, 152].

124   Taylor 1998a (n. 20 above), 297.

between anthropocentrism, individual freedom and property rights on the one hand, and global environmental degradation on the other.[125] And equally it can be argued that there is a causal link between anthropocentrism and territorial sovereignty on the one hand, and global environmental degradation on the other.[126]

The Transformational Approach

The attempts to incorporate global environmental concerns into existing instruments and principles discussed so far all have their merits. They strengthen and sharpen the expectations towards state responsibility, treaty-making, shared responsibility, environmental rights and human rights. All those expectations reflect an increasingly held view that state sovereignty may be more of an obstacle to, rather than a vehicle for, global environmental protection.

Building upon existing environmental instruments and principles, however, does not go far enough. The reformist approach ignores the causes leading to the environmental crisis in the first place. There is little point relying on territorial sovereignty for solutions, if the dogma is part of the problem. Historically, the modern state grew out of concerns for property and territorial rights and has always maintained its role for their protection.[127] On the other hand, the means of this protection, state sovereignty, has proven to be a 'myth'[128] as it is incapable of meeting those concerns. Against this background, the reformist approach towards environmental governance appears less realistic than the transformational approach of reconsidering territorial sovereignty.

Principally, 'reconsidering' includes the option of abolishing the idea of state sovereignty altogether. After all, this option has been pursued by many theorists throughout the twentieth century.[129] Given the right circumstances, even states themselves may, gradually, transfer their sovereignty to a new supranational level.[130] Also, states have become accustomed to restrictions upon their sovereignty in the form of international human rights. Precisely these restrictions, however, assume the continued existence of the sovereign state. The state is not only violator, but also protector of human rights. The anarchical characteristics of present economic and financial globalization may, in fact, require a better utilization of remaining state power. State sovereignty remains an important concept.

---

125   Bosselmann 1995 (n. 6 above), 51–62, 226–30.

126   On the historical links between property rights and the concept of territorial sovereignty see Bosselmann 1995 (n. 6 above), 51–73; Taylor 1998a (n. 20 above), 118–22.

127   Bosselmann 1995 (n. 6 above), 51–73; Taylor 1998a (n. 20 above).

128   Bosselmann 1995 (n. 6 above), 74.

129   For example, Politis, N. (1925/I), 'Le probléme des limitations de la souveraineté et la théorie de l'abus des droits dans les rapports internationaux', *Recueil des Cours à l'Académie de Droit International* 6, 1–121, 18; Korowicz, M. (1961/I), 'Some Present Aspects of Sovereignty in International Law', *Recueil des Cours à l'Académie de Droit International* 102, 1–120, 16; Jenks, C.W. (1969), *New World of Law?* (London, Longmans Green).

130   Many commentators believe that the member states of the European Union have either lost, or are about to lose, their sovereignty; e.g. MacCormick, N. (1993), 'Beyond the Sovereign State', *Modern Law Review* 56, 1–18.

The transformational approach as, for example, advocated by Richard Falk,[131] Günther Handl[132] or Lynton Caldwell,[133] assumes the existence of state sovereignty, but relates it to the idea of civil society. The perspective shifts from state-centredness to globality and sets a new agenda. In strategic terms, the functions of state sovereignty are relative to the needs of civil society. In political terms, states are not expected to relinquish sovereignty by means of sweeping changes, but to adjust sovereignty to global realities.

A redefinition of territorial sovereignty is based on the assumption that the environment is not territorial, but global. This means that environmental governance needs to be distinguished from state governance. In institutional and legal terms, however, environmental governance needs to respond to state governance as the conventional form of governance. Two steps may be required to allow environmental governance to find its place within state sovereignty.

A first step would be to see the parallels between property and sovereignty as indicated earlier. Just as private property cannot be defined without its social dimensions,[134] state sovereignty cannot be defined without its international dimensions.[135] Both are neither absolute, nor independent from the system they are operating in. The international dimension of sovereignty could, therefore, be perceived as an integral part of it.[136] This conceptual limitation to state sovereignty serves the protection of the international community of states.

As mentioned earlier, the community of states not the same as the community of ecosystems. The difference is important to note as, traditionally, the community of states has defined any limitations to sovereignty on the basis of mutual expectation. Such reciprocity cannot be expected with respect to the global environment. If the protection of the global environment is determined through the reciprocal expectations of states, states interests *in* the environment will again prevail over interests *of* the environment. Global environmental protection would remain derivative and secondary.[137] The recognition of the collective dimension of sovereignty can, therefore, only be the first step.

The second step is to add the ecological dimension to concepts of property and sovereignty, respectively. Just as property is determined by ecological realities, so

---

131   Falk, R. (1989), *Revitalizing International Law* (Iowa, Iowa State University Press).

132   Handl, G. (1991), 'Environmental Security and Global Change: The Challenge to International Law', in Lang, W. et al, (eds), *Environmental Protection and International Law* (London, Graham & Trotman/Martinus Nijhoff), 59.

133   Caldwell, L. (1984), *International Environmental Policy: Emergence and Dimensions* (Durham, NC, Duke University Press).

134   As expressed, for example, in continental European constitutions.

135   See Simma B. (1993), 'Does the UN-Charter Provide an Adequate Legal Basis for Individual or Collective Responses to Violations of Obligations erga omnes?', in Delbrück, J. (ed.), *The Future of International Law Enforcement* (Berlin, Duncker & Humblot), 125, 129.

136   See Allott, P. (1990), *Eunomia. New Order for a New World* (New York, Oxford University Press), 296; Bernhardt, R. (1976), 'Ungeschriebenes Völkerrecht', *Zeitschrift für ausländisches öffentliches Recht und Völkerrecht* 36, 50–76, 58.

137   Odendahl 1998 (n. 75 above), 352; see also Bosselmann 2004 (n. 75 above), 312.

is sovereignty. The crucial question is, therefore, whether an ecological dimension should be part of the definition of property[138] and of sovereignty.

The justification for taking this step is the indivisibility of the global environment. To accommodate environmental indivisibility in international law, two approaches are possible. Either the idea of territorial sovereignty is dismissed altogether on the grounds that the environment cannot be divided up into territories,[139] or territorial sovereignty is redefined on the grounds that there are use and conservation aspects of the (indivisible) environment to be dealt with. Following this approach, territorial sovereignty needs to recognize the dual nature of the environment it occupies. The dual nature is best understood as consisting of a right to use 'territorial' natural resources and an obligation to protect the environment. Territorial sovereignty over natural resources (or 'fruits' of the environment) can only be perceived, therefore, as inherently linked to an obligation to protect the environment ('the substance').

Sovereignty restricted in such a principal way contains the right to use, but not to abuse, pollute, exploit or otherwise overuse the environment. In practical terms, the difference between use and abuse of the environment may be difficult to detect, but a discernible difference exists nevertheless. If the (legal) use of the environment has effects that could threaten the integrity of ecosystems it can be seen as (illegal) abuse. For example, extraction of minerals may be a legitimate exercise of territorial sovereignty, but associated environmental effects (erosion, loss of vegetation and biodiversity) may not; fertilization of soils may be a legitimate exercise of territorial sovereignty, but causing threats to their fertility may not; forestry may be legitimate, but threats to biodiversity may not; fishing in coastal waters may be legitimate, over-fishing may not; use of fossil fuels may be legitimate, contributing to the greenhouse effect may not. In positive terms, the use of resources is covered by territorial sovereignty only within the parameters of ecological sustainability. As a result, the current 'sovereign right of states to exploit their resources pursuant to their own environmental policy' (Principle 21 Stockholm Declaration and Principle 2 Rio Declaration) converts to the new sovereign right of states to exploit their resources pursuant to the principle of sustainability.

The reference to the principle of sustainability is crucial to the redefinition of territorial sovereignty. There is little merit in restricting territorial sovereignty without noting the integrity of the Earth's ecosystems. If the restriction depends on 'environmental components' of 'global importance' and 'the consequences of their

---

138   As, for example, developed in Bosselmann, K. (1998), *Ökologische Grundrechte*, 100–124; Taylor, P. (1998b), 'From Environmental to Ecological Human Rights: A New Dynamic in International Law?', *Georgetown International Environmental Law Review* 10, 384–85; and Schröter, M. (1999), *Mensch, Erde, Recht. Grundfragen ökologischer Rechtstheorie* (Baden-Baden, Nomos), 246–53; See also Hunter, D.B. (1988), 'An Ecological Perspective of Property: A Call for Juridical Protection of the Public's Interest in Environmentally Critical Resources', *Harvard Environmental Law Review* 12, 311; Rieser, A. (1991), 'Ecological Preservation as a Public Property Right: An Emerging Doctrine in Search of a Theory', *Harvard Environmental Law Review* 15, 393.

139   Such approach may be accused of 'throwing the baby out with the bath water'. The total abolition of territorial sovereignty, even if conceivable, may render render states (even more) defenceless against interventions from states, terrorists, multinational companies, etc.

potential degradation or destruction for all',[140] anthropocentric and economic state interests will dictate the content of territorial sovereignty. Not all 'environmental components', but only those of assumed 'global importance' would be exempted from the sovereign right of exploitation. This would continue the transboundary approach to pollution, assuming environmental boundaries between national territory and areas outside. Effects limited to national jurisdiction would be seen as covered by territorial sovereignty and transboundary effects as not covered.[141] The ecological reality does not suggest such distinction. Whether or not intra-territorial activities have global effects is not determined by their 'global importance'. For example, clear-felling of a very small forest might be seen as intra-territorial and not transboundary, thus covered by territorial sovereignty. The cumulative effects of local activities may, however, be of global significance regardless of how insignificant they appear when judged on their own. It is, therefore, wrong to distinguish between 'environmental components' of 'global importance' and those that are not.[142] The focus should be on activities and on the distinction between activities that are ecologically sustainable and those that are not.

The idea that states should act as trustees or guardians underlies both approaches, the sustainability governance approach and the state governance approach, to restricted territorial sovereignty. It makes, however, a difference whether the trusteeship role is derived from the environment *per se* or from the community of states. This second form of trusteeship would be limited to environmental components of global importance or global commons.[143] The first form would include the entire environment within territorial jurisdiction. States are trustees of the entire environment, not just of globally important 'components'.[144] Apart from the ecological importance involved here, there is also a global justice issue to be considered. Poor states with rich natural resources ('developing' countries) might be more restricted in their sovereignty than rich states with poor natural resources ('developed' countries), if trustee functions are limited to globally significant resources.

Governance for sustainability requires the acceptance that the environment is entrusted to the individual state not by virtue of its sovereignty or any other form of legal entitlement, but by virtue of the laws of physics: any territory exists in an indivisible global environment. It follows that the environment belongs neither to states nor to humanity, but only to itself due to its intrinsic value. States, therefore, cannot claim sovereignty or ownership over the environment. The environment is a privilege, not a right, and any entitlements are limited to the sustainable use of the environment's resources.

---

140   Kiss 1999 (n. 44 above), 150.

141   See for example, Shrijver 1997 (n. 15 above), 290–92.

142   US American public trust doctrines, although helpful, carry this inherent flaw. See for example, Taylor 1998b (n. 138 above), 386–92; Hunter 1988 (n. 138 above), 317–19, 375–6.

143   'States having under their jurisdiction such (i.e. globally important) environmental components should be considered as trustees in charge of their conservation'; Kiss 1999 (n. 44 above), 150.

144   Taylor 1998b (n. 138 above), 394; Odendahl 1998 (n. 75 above), 353, 363; Bosselmann 1995 (n. 6 above), 259.

Sustainability governance conceptually restricts territorial sovereignty leading to a paradigm shift in international environmental law: not state sovereignty setting limits to environmental protection, but environmental protection setting limits to state sovereignty. The common interest would, perhaps for the first time in the history of international law, gain control over its own creation, i.e. state sovereignty.

Implementation: The State as Trustee

The incorporation of ecological sustainability into the concept of territorial sovereignty creates a fundamental duty to protect the integrity of Earth's ecosystems. This duty would exist independently of treaty obligations or any other consensus-building process in international law. It would exist by virtue of being a sovereign state. No state or legally recognized territorial entity can claim sovereign rights over natural resources without accepting the duty to use them sustainably. The state is owner of resources and, at the same time, trustee or guardian of the environment.

The general idea of environmental trusteeship is by no means new. The history of public trusts goes back to the Roman law concept of *res communis*, according to which environmental resources such as air, water and the sea are for common use. Public trust thinking has led to the concept of common property in English and American common law and in the German *Allmende* system.

For the modern development of US environmental law in the early 1970s, Joseph Sax rediscovered the effectiveness of public trusts for environmental resources.[145] Later he called for 'liberating the public trust doctrine from its historic shackles'[146] by observing that democratic decision-making often fails to recognize the ecological significance of natural resources. The objective of public trusts must be not only to ensure common use of natural resources, but also the preservation of ecological integrity. A number of commentators have noted that the actual application of the public trust doctrine in US environmental law falls short in this regard.[147] The (anthropocentric) common use purpose competes with the (ecocentric) preservation purpose of genuine trusteeship, but only the former is being addressed in legislation.[148] Prue Taylor criticizes the anthropocentric limitations of property rights and public trust doctrines in the US as fundamentally undermining any long-term prospect for protecting ecological integrity. That is why we need an ecological approach to both property rights and public trusteeship.[149]

The concept of public trusteeship for environmental resources exists in environmental laws of the USA, Canada, South Africa, India, the Philippines,

---

145 Sax, J.L. (1970), 'The Public Trust Doctrine in Natural Resources Law: Effective Judicial Intervention', *Michigan Law Review* 68, 471–556.

146 Sax, J.L. (1980), 'Liberating the Public Trust Doctrine from its Historic Shackles', *UC Davis Law Review* 14, 185–94.

147 For example Lazarus, R.J. (1986), 'Changing Conceptions of Property and Sovereignty on Natural Resources: Questioning the Public Trust Doctrine', *Iowa Law Review* 71, 631–716; Wilkinson, C.F. (1989), 'The Headwaters of the Public Trust: Some Thoughts on the Source and Scope of the Traditional Doctrine', *Environmental Law* 19, 425–72.

148 Hunter 1988 (n. 138 above); Rieser 1991 (n. 138 above).

149 Taylor 1998b (n. 138 above), 384–92.

Australia[150] and New Zealand. It is less developed within Europe's civil law system, but examples for environmental trust agencies can be found in Sweden and Italy.[151]

As far as international law is concerned, there is a long-established recognition of 'shared resources', such as fresh water and international watercourses,[152] certain species[153] or *res communis*, expressed in common heritage doctrines.[154] In addition, the concept of common concern has a long tradition in international environmental treaties, for example, with respect to whaling, Antarctica, endangered species,[155] biological diversity and climate change.[156] The protection of fundamental values is a common concern of any community including local and national communities. Alexander Kiss has stressed that this is also true for the international community. Maintaining international peace and security, improving health, food supply and education, the protection of human rights, etc. all reflect common concerns of humanity, whether being expressed as such or merely assumed in treaties, customary law or soft law. And while the common concern is a general concept without specific rules and obligations, it is the basis for the international community to act. Thus, Kiss interprets the invocation of common concern in key environmental treaties such as the Biodiversity Convention or Climate Change Convention as proclaiming that the global environment is a matter of common concern for humanity.[157] This means that actions affecting it 'are no longer solely within the domestic jurisdiction of states'.[158] One application of this view is, for example, that states are free to negotiate global trade agreements, but not at the expense of the protection of ecological integrity: 'what is needed is to adapt international trade and the functioning of the global economy to the requirements of the preservation of human health and of the global environment – and not the contrary'.[159] If, and in so far as the global environment is a matter of common concern, the question arises as to how it can be related to the concept of environmental trusteeship.

Environmental trusteeship is in itself an almost 'common concern' among international environmental lawyers. There is an abundance of ideas. As Peter Sand

---

150   Sand, P. (2006), 'Global Environmental Change and the Nation State: Sovereignty Bounded?', in Winter, G. (ed.), *Multilevel Governance of Global Environmental Change* (Cambridge, Cambridge University Press), 519–38, 523–4.

151   Ibid. 524.

152   Kiss and Shelton 2004 (n. 23 above), 122–5, 399–402, 407–19.

153   The Preamble to the 1971 Convention on Wetlands ('Ramsar') recognizes 'that waterfowl in their seasonal migrations may transcend frontiers and so should be regarded as an international resource'.

154   Kiss and Shelton 2004 (n. 23 above), 249–50.

155   The Preamble to the 1979 Migratory Species Convention recognizes 'that wild animals in their innumerable forms are an irreplaceable part of the earth's natural system which must be conserved for the good of mankind'.

156   Kiss and Shelton 2004 (n. 23 above), 250–54.

157   Ibid. 251.

158   Ibid.

159   Kiss 1999 (n. 44 above), 143–53, 153.

points out,[160] various forms of 'trusteeship', 'guardianship', 'custodianship' or 'stewardship' status have been suggested for a range of different environments. They include the marine environment in coastal waters and exclusive economic zones,[161] marine resources in specific regional seas such as the Mediterranean[162] and the South Pacific[163], living ocean resources in general,[164] Antarctica,[165] the rain forests in Latin

160   Sand, P. 2006, (n. 150 above), 528–9; see also Sand, P. (2004), 'Sovereignty Bounded: Public Trusteeship for Common Pool Resources?', *Global Environmental Politics* 4:1, 47–71, 52–3.

161   Beesley, J.A. (1973), 'The Canadian Approach to International Environmental Law', *Canadian Yearbook of International Law* 11, 3–12; Archer, J.H. and Jarman, C. (1992), 'Sovereign Rights and Responsibilities: Applying Public Trust Principles to the Management of EEZ Space and Resources', *Journal of Ocean and Shoreline Management* 17, 251–64; Hildreth, R.G. (1993), 'The Public Trust Doctrine and Coastal and Ocean Resources Management', *Journal of Environmental Law and Litigation* 8, 221–36.

162   Raftopoulos, E. (1992), 'The Barcelona Convention System for the Protection of the Mediterranean Sea against Pollution: An International Trust at Work', *International Journal of Estuarine and Coastal Law* 7, 27–41.

163   Fong, G. (1993), 'Governance and Stewardship of the Living Resources: The Work of the South Pacific Forum Fisheries Agency', in Van Dyke, J.M., Zaelke, D. and Hewison, G. (eds), *Freedom for the Seas in the 21st Century: Ocean Governance and Environmental Harmony* (Washington, DC, Island Press), 131–41.

164   Van Dyke, J.M. (1993), International Governance and Stewardship of the High Seas and Its Resources, in Van Dyke, J.M., Zaelke, D. and Hewison, G. (eds), *Freedom for the Seas in the 21st Century: Ocean Governance and Environmental Harmony* (Washington, DC, Island Press),13–22; Zahren, W.M. von (1998), 'Ocean Ecosystem Stewardship', *William and Mary Environmental Law and Policy Review* 23, 108–20.

165   Suter, K. (1991), *Antarctica: Private Property or Public Heritage?* (London, Pluto Press), 170.

America,[166] biological resources generally,[167] the oceans,[168] the global atmosphere,[169] the global commons[170] or the entire global environment.[171]

As part of the UN reform, the then UN Secretary-General Kofi Annan proposed that the United Nations Trusteeship Council:

> be reconstituted as the forum through which Member States exercise their collective trusteeship for the integrity of the global environment and common areas such as the oceans, atmosphere and outer space. At the same time, it should serve to link the United Nations and civil society in addressing these areas of global concern, which require the active contribution of public, private and voluntary sectors.[172]

This proposal includes not only common areas, but also the integrity of the global environment. Accordingly, states would act as trustees also with respect to their 'internal' environment. Legally, this is quite conceivable. One example is the designation of protected areas under the World Heritage Convention,[173] effectively restricting the permanent sovereignty over national resources.[174] Another example is

---

166   Tarlock, A.D. (1997), 'Exclusive Sovereignty versus Sustainable Development of a Shared Resource: The Dilemma of Latin American Rainforest Management', *Texas International Law Journal* 32, 37–66.

167   Gebel, T., 'Der Treuhandgedanke und die Bewahrung der biologischen Vielfalt: Einschränkung der territorialen Souveränität durch treuhänderische Verwaltung von lebenden Umwelt-Ressourcen', Ph.D. dissertation (Pro Universitate, Sinzheim, 1998); Foundation on Economic Trends (2001), *Draft Treaty to Share the Genetic Commons* (Washington, DC, FOET; repr. in South Letter 2:37), 29–30.

168   Sand, P. (2007), 'Public Trusteeship for the Oceans', in Wolfrum, R., Ndiaye, T.M. and Kojima, C. (eds), *Law of the Sea, Environmental Law and Settlement of Disputes: Liber Amicorum Judge Thomas A. Mensah* (Leiden, Martinus Nijhoff).

169   Taylor 1998a (n. 20 above), 283.

170   Bosselmann 1992 (n. 4 above), 301–405; Stone, C.D. (1993), *The Gnat is Older than Man: Global Environment and Human Agenda* (Princeton, NJ, Princeton University Press), 83; Cleveland, H. (1993), 'The Global Commons: A Global Commons Trusteeship Commission Is Needed to Guide our Use of the Oceans, Antarctica, the Atmosphere, and Outer Space', *The Futurist*, May–June, 9–13.

171   Kiss, A. (1989), *Droit international de l'environnement* (Paris, Pedone), 19; Taylor 1998a (n. 20 above), 300–305; Odendahl 1998 (n. 75 above); Brown, P.G. (2001), *The Commonwealth of Life: A Treatise on Stewardship Economics* (Montréal, Black Rose Books), 88–91, 121–37; Bosselmann, K. (2004), 'Environmental Governance: A New Approach to Territorial Sovereignty', in Goldstein, R.J. (ed.), *Environmental Ethics and Law* (Aldershot, Ashgate), 293–313.

172   Report of the UN Secretary-General to the General Assembly, 'Renewing the United Nations: A Programme for Reform' (14 July 1997; UN Doc. A/51/950).

173   UNESCO Convention for the Protection of the World Cultural and Natural Heritage (16 November 1972, 11 ILM 1358).

174   Michael Bothe comments: 'It appears that national sovereignty is modified by the principle of common entitlement' as a '*droit de regard* concerning national measures of preservation'; Bothe, M. (2006), 'Whose Environment? Concepts of Commonality in International Environmental Law', in Winter, G. (ed.), *Multilevel Governance of Global Environmental Change* (Cambridge, Cambridge University Press), 539–58, 549.

the sovereign rights of coastal states in their marine exclusive economic zone that are qualified by certain obligations owed to the international community (Articles 61–70 of the UN Convention on the Law of the Sea). Further, sovereign rights over access to genetic resources *in situ* are complemented by obligations to allow access for other parties (Article 1(2) of the Convention on Biological Diversity). Finally, some conventions on nature conservation require states to recognize, for example, domestic 'waterfowl ... as an international resource' (Preamble to the Ramsar Convention) and 'wild animals [as] ... part of the earth's natural system which must be conserved for the good of mankind' (Preamble to the Migratory Species Convention). In all instances, treaty law has imposed certain fiduciary duties on parties within their territory. Michael Bothe interprets them as recognition of a '*droit de regard*' of other states or of the organs of a treaty regime'. He concludes: 'The principle of state sovereignty is maintained. But the sovereignty is limited by these Conventions, yet not to an extent that the regime could be compared to a trust or similar concept of private law'.[175]

While this is certainly true, the examples illustrate the principal possibility of redefining territorial sovereignty through the incorporation of fiduciary obligations. State sovereignty is not at stake. It is also clear that the current spectrum of environmental principles, from the duty to cooperate, common concern, common heritage, common but differentiated responsibilities, to the precautionary principle and sustainable development, is insufficient and too weak to support the idea of a general environmental trusteeship of the state.

Conclusion

Asking the sovereign state to accept a fundamental fiduciary duty is a bit like asking the fox to look after the chicken. The appetite of states to exploit the global environment is greater than their interest to preserve it. On the other hand, this logic is increasingly becoming the 'logic of self-determination'[176] not just of states, but of humanity at large. International law needs to accept the preservation of the Earth's integrity as the foremost 'common concern of humanity',[177] simply because we must.

Ecological sustainability is the only principle that could reverse the current logic. Like the precautionary principle, the principle of sustainability has its roots in ecological insights rather than legal anthropocentrism. And it is not compromised in a way that the concept of sustainable development currently is. If we accept, therefore, that the integrity of the Earth's ecosystems cannot be sliced up in pieces that fit into areas inside or outside national boundaries, then states need to be bound by a universal principle. Like the universal principles of justice and human rights, sustainability would define the internal as well as external functions of the state.

---

175   Ibid. 551.

176   Bosselmann 1995 (n. 6 above), 40.

177   'The global environment is interdependent and constitutes an ecological unit, making its protection a matter of common concern for all humanity'; Kiss and Shelton 2004 (n. 23 above), 251.

Internally, the state's trustee role involves a fundamental obligation to protect the environment for its own sake. Such a state obligation could be established through constitutional amendments following the examples in the constitutions of Germany (Article 20a), Austria (Articles 10–12), Switzerland (Article 24), Netherlands (Article 21), Sweden (Article 2), Finland (Article 20) and Greece (Article 24).[178]

Externally, states have a number of options available including the adoption of the Earth Charter, the negotiation of a global convention on sustainable development, the creation of customary law or the drafting of soft law documents. As we have seen, a certain degree of environmental trusteeship is already part of international law. However, a more promising prospect is the ever-increasing importance of global public opinion. Public pressures grow from within and outside the state and can, for example, lead to interesting judicial law development.[179] But even without new case law we can expect major legal changes as global civil society and ecological citizenship gain strength. In the following final chapter we will explore this strength and its implications further.

If the new role of the state as environmental trustee seems utopian, the rationale behind it is certainly more reality-based than the rationale behind the current concept of territorial sovereignty. As sovereign *and* trustee of the environment, the state would not lose its sovereignty, but foster it in a moral and factual sense.

---

178   Above, 124–125.

179   See for example, Judge Weeramantry's reference to a 'principle of trusteeship for earth resources'; *Case Concerning the Gabčikovo-Nagymaros Project* (*Hungary/Slovakia*), 1997 ICJ, 37 ILM 162 (1998), Separate Opinion of Judge Weeramantry, 204, 213.

Chapter 6

Governance for Sustainability

Introduction

Like the previous chapter, this final chapter is concerned with institutions of governance. The state is a central institution of environmental governance. Relating the principle of sustainability to the concept of territorial sovereignty has led us to include trusteeship in the state's functions. In a similar vein we can relate the principle of sustainability to other institutions participating in global environmental governance. They include international organizations (IGOs), non-governmental organizations (NGOs) and civil society with the idea of citizenship at its core. Together with states they all participate in environmental governance.

The term 'environmental governance' is used loosely here to include the various institutions and structures of authority engaged in the protection of the natural environment. However, when we assess the actual performance of environmental governance and ask how successful it has been, the term itself becomes political. 'There is little dispute that better governance is required', says Lorraine Elliott, but 'a precise definition of what this means or what it requires is elusive'.[1] It is helpful to contrast environmental governance with governance for sustainability, both representing very different concepts. Governance for sustainability is value-based, acknowledging the fundamental importance of the preservation of Earth's ecological integrity.

The purpose of this chapter is to describe how the principle of sustainability can inform the system of environmental governance, and how existing institutions may be able to respond. Following a definition of 'governance for sustainability', we will outline the new governance framework proposed by the Earth Charter, then examine several levels of governance, i.e. the global level, the regional/transnational level and the level of civil society (acting locally, nationally and globally). The aim is to show some dramatic changes in the way multi-level governance is operating today and to highlight citizenship and civil society as catalysts for change.

What is Governance for Sustainability?

Governance for sustainability is fundamentally different from environmental governance. In most 'modern' societies environmental governance has been a minor concern, an add-on, or a minimalist, shallow programme, designed to stay at the

---

1   Elliott, L. (2004), *The Global Politics of the Environment*, 2nd edn (New York, Palgrave Macmillan), 94.

periphery of public governance.² It is the poor cousin of economic governance (for ongoing growth in productivity and profit). Whereas governance for sustainability has its origins in holistic awareness and responsible values, the current emphasis on economic governance is the product of values that place personal, short-term gain over social equality and human security. The problem is not so much the underlying economic rationality, but its dominance in any form of governance.³ The defensive, reactive, expert-based, problem-solving focus of environmental governance contrasts with our need for imaginative, proactive design and redesign approaches to planetary health and well-being.

Essentially, we need to think differently about governance and people's role in it. Governance can no longer be limited to purely social relationships. We need also to reflect our ecological relationships. The traditional focal point of governance is the human community. The new focal point needs to be the wider community of life. The inclusion of all life (in addition to human life) marks an important shift.

Shifting our understanding of the purpose of governance systems is at the heart of the sustainability approach. The shift from traditional anthropocentrism to ecocentrism may be revolutionary in its conceptualization. However, making the shift in practice is likely to be more evolutionary and gradual. For example, in a national legal system the purpose of an Act may be reflected in the long title or objects. The inclusion of the sustainability principle in the purpose description of the New Zealand Resource Management Act 1991 had opened new doors as 'it provides for sustainability to be the overriding objective, comprehending sustainability of not only humankind but also ecology and biodiversity in all aspects'.⁴ Legislation may also stipulate that in making a particular decision, such as whether or not to grant a permit, a decision-maker must apply certain principles (for example, the precautionary principle) and must have regard to certain matters, such as the views of interested and affected parties. For example, legislation commonly refers to decisions that must be taken in the public or national interest. These can be replaced by new terms such as 'in the interests of the whole community' which can then be defined to mean the long-term, collective interests of present and future generations of humans and other species. Similarly, it might be appropriate to require land-use planning decisions to be based on an assessment of what would constitute a wise use of land from the perspective of the whole community, rather than an assessment of (human) 'needs and desirability' as is so often the case in planning decisions.

The sustainability approach to governance is not intended to be a lofty, transcendental theory. It must be, literally and metaphorically, down to Earth, and grounded in our experience of the natural world. At every turn we must look at

---

2 On the history of environmental governance see Bosselmann, K. (1995), *When Two Worlds Collide* (Auckland, RSVP), 51–73.

3 'The problem with economic man is not that he is an imaginary theoretical fiction, but that there are too many of him. Economic rationality, which is supposedly a purely formal assumption, has bled out into the community and has become a foundation for ethics': Emerson, J., 'What does Economic Rationality Do?' <http://www.idiocentrism.com/economic%20rationality.htm>, accessed 1 December 2007.

4 Nolan, D. (2005), *Environmental and Resource Management Law*, 3rd edn (Wellington, LexisNexis), 92.

particular laws and aspects of our governance systems and ask ourselves 'How would this look from the perspective of the whole Earth community?' What might a particular law say if the subjects of it were not only human beings and we really acted as if the flourishing of the whole Earth community was our primary concern? We can ask, for example, how would we reformulate a requirement that interested and affected parties must be given an effective right to participate in the making of decisions that impact upon their environment, if we accept that humans are not the only interested and affected parties? Do we have adequate techniques and methodologies for 'consulting' and ascertaining the current and future interests of rivers? If not, how might they be developed?

Questions like these signal a sense of urgency and focus, but they also reflect the ecological orientation that separates sustainability governance from outmoded environmental governance.

The Vision of the Earth Charter

The leading text of environmental law in New Zealand considers the Earth Charter to be the inspiration for a governance model based on sustainability. Referring to the 'importance of non-human ecology or broad biodiversity' and 'earth guardianship and trusteeship principles',[5] it introduces the values and principles of the Earth Charter and stresses that 'the law applicable in New Zealand recognises the intrinsic value of ecosystems' and 'seeks to provide a framework to resolve the conflicting values and objectives'.[6] Such linking of the Earth Charter to a national legal system may not be typical for environmental law texts; it does, however, demonstrate the synergies between ethics, law and governance.

Sustainability, as we have seen, is a fundamental ethical principle with clear direction for the design of law and governance. And nowhere is this direction more clearly expressed than in the Earth Charter. The Earth Charter is the foundational document for 'building a just, sustainable, and peaceful global society in the 21st century'.[7] The emphasis here is on 'building', i.e. the principles and institutions creating governance for sustainability. What institutional changes does the Earth Charter promote and how can they be understood in terms of global ecological governance?

The ambition of the Earth Charter is 'to bring forth a sustainable global society founded on respect for nature, universal human rights, economic justice, and a culture of peace' (Preamble, first paragraph). The Charter itself organizes its 77 principles around four main themes: 'Respect and Care for the Community of Life' (Principles 1–4), 'Ecological Integrity' (5–8), 'Social and Economic Justice' (9–12) and 'Democracy, Nonviolence, and Peace' (13–16).

The first two themes contain the ecological principles, the other two themes contain the social and economic principles of a sustainable global society. We can

---

5   Ibid. 29.

6   Ibid. 28.

7   The Earth Charter Initiative, <www.earthcharter.org>, accessed 1 December 2007.

say, therefore, that the Earth Charter reflects the concept of sustainable development with its three pillars of environmental, social and economic equity. This three-pillar model is commonly accepted by states, expressed in many soft law documents[8] and promoted within business.

The Earth Charter, however, takes a crucial further step. It not only defines the three pillars, it also organizes them in a particular way. Environmental concerns are perceived differently from social and economic concerns. The environment is not perceived as the resource base for human consumption and not as one of three equally important factors, but as the basis of all life. This shift from a narrow human-centred to a broader life-centred perspective is expressed in respect and care for the community of life and ecological integrity as the two overarching principles of governance.

The ethics of sustainability governance is also spelled out in the Preamble: 'It is imperative that we, the peoples of Earth, declare our responsibility to one another, to the greater community of life, and to future generations'. The new element that the Earth Charter promotes, in contrast to any other international document, is our responsibility to the greater community of life. In jurisprudential terms, such three-fold moral responsibility translates to a three-dimensional concept of (ecological) justice[9] and ecological human rights.[10]

However, in our context we are less concerned with the normative foundations of sustainability governance than with the actors and institutions promoting them. As these normative foundations are obviously different from those of traditional forms of global governance, it could be asked whether they require different advocates as well. Given the systemic character of ecological destruction and the persistence of unsustainable forces, we may still believe in humanity's *general* ability to learn, but cannot rely on corporate and governmental institutions in their present form. The new driving forces – people and ideas inside and outside these institutions – need to be informed by a strong sense of ethics.

From a legal perspective, the most relevant aspects are the relevant institutions. A lawyer can ask what functions institutions such as citizenship, civil society, states and international organizations have for the formation of sustainability governance. We can distinguish between two kinds of institutions. The first kind is the concept of citizenship in both its manifestations, individual citizenship and civil society. The second kind are the institutions formed by them. There is an important role for civil society in advocating trusteeship functions of the state, as we have seen. There is also a need to more actively engage with international organizations. The relationship between civil society and governments is often perceived as a dialogue of partners.[11] The reality is rather different, of course, but does not necessarily require a 'bottom-

8   For example, Rio Declaration on Environment and Development (the Rio Declaration) (Rio de Janeiro, 13 June 1992; UN Doc. A/Conf.151/26 (vol. I); 31 ILM 874 (1992)); and Johannesburg Declaration (Johannesburg, 4 September 2002, UN Doc. A/Conf.199/20).

9   See Chapter 3.

10   See Chapter 4; see also Bosselmann, K. (2008), 'The Way Forward: Governance for Ecological Integrity', in Westra, L., Bosselmann, K. and Westra, R. (eds), *Reconciling Human Existence and Ecological Integrity* (London, Earthscan), 319.

11   Especially under the sustainable development agenda since 1992.

up' approach for a remedy. The relationship between civil society and state probably works in a dialectic manner with 'top-down' and 'bottom-up' processes operating simultaneously.

Global governance can be perceived in many shapes and forms. However, any realistic concept would include states. States are at the centre of the current architecture of international governance. Thus, the question is not whether, but how states can contribute towards effective global governance. Equally, international organizations such as the United Nations are not unsuitable *per se*. Rather, they will reflect the ability of states to change.

The Earth Charter is decidedly inclusive. It refers to citizens, civil society, business, nations and governments as all having to work together. Here are some characterizations starting with citizens.

The Preamble proclaims: 'We are at once *citizens*[12] of different nations and of one world in which the local and the global are linked. Everyone shares responsibility for the present and future well-being of the human family and the larger living world'. It then continues: 'The emergence of *global civil society* is creating new opportunities to build a democratic and humane world. Our environmental, economic, political, social, and spiritual challenges are interconnected, and together we can forge inclusive solutions'. The citizenship concept expressed here is for citizens to be aware of their global and ecological responsibilities. At the same time, the legitimacy of corporate and governmental institutions is assumed. Global civil society is not diametrically opposed, but complementary to established institutions. This is visible in the Preamble's plea for a 'global partnership to care for the Earth' and in the concluding statement ('The Way Forward'): 'The *partnership of government, civil society, and business* is essential for effective governance'. And: 'In order to build a sustainable global community, the *nations* of the world must renew their commitment to the United Nations, fulfil their obligations under existing international agreements and support the implementation of the Earth Charter principles'.

On the basis of these statements, we can examine the institutions responsible for sustainability governance. We will first look at international organizations created by states and their potential for advocating change. The innovative element of these organizations is (emerging) global civil society with its core idea of ecological citizenship.

International Organizations

Traditionally, the global view on governance is expressed by international organizations. However, the existing intergovernmental institutions are inadequate to deal with the complex, integrated, interdependent and, most importantly, political nature of environmental problems. Already in 1987 the Brundtland Report noted that:

---

12   Emphasis added.

institutions tend to be independent, fragmented and working to relatively narrow mandates with closed decision processes. The real world of interlocked economic and ecological systems will not change; the policies and institutions concerned must.[13]

As long as states have the monopoly for determining the role of international institutions, they will follow their needs, not the needs of ecological governance. As states favour short-term economic objectives over long-term environmental goals, they 'do not necessarily choose the tools which are most *effective* in achieving the policy goal, rather they will choose tools which will benefit them most politically'.[14] It is crucial, therefore, to perceive states as dynamic organizations capable of learning and adopting a trusteeship attitude to the global environment.

The United Nations System

The following section discusses some reform developments within the United Nations (UN) system.

Since it was founded in 1972, the United Nations Environment Programme (UNEP) has been a cornerstone of global governance for the environment. Recently, however, the organization has been called into question due to the increasing diversity of international organizations acquiring environmental responsibility and the perceived lack of effectiveness of UNEP.[15] UNEP does not possess executive powers; instead, its primary function is to monitor and coordinate environmental governance, which includes engaging in partnerships with other intergovernmental and non-governmental organizations.[16] Perceived ineffectiveness of UNEP has been linked primarily to the organization's geographical isolation in the UN system, an insufficient mandate, lack of support from governments and a low budget.[17] Essentially, critics argue that UNEP is too small, too poor and too remote to coordinate and promote sustainability effectively.[18] This is particularly so as the international system of environmental governance is becoming ever more decentralized and other international organizations targeting environmental governance often have 'better funding ... clearer and stronger mandates, and greater support'.[19]

---

13   World Commission on Environment and Development (ed.) (1987), *Our Common Future*, 'Brundtland Report' (Oxford, Oxford University Press), 310.

14   Wynter, M. (1998), 'The Compatibility of Trade and Environmental Concerns Lessons From the Shrimp–Turtle Dispute', in Anghie, A. and Sturgess, G. (eds), *Legal Visions of the 21st Century: Essays in Honour of Judge Christopher Weeramantry* (The Hague and Boston, Kluwer Law International), 6.

15   See for example, Elliott 2004 (n. 1 above), 102.

16   Ibid. 97.

17   Ibid. 98.

18   For a summary of these critiques, see for example Downie, D.L. and Levy, M.A. (2000),'The United Nations Environment Program at a Turning Point', in Chasek, P. (ed.), *The Global Environment in the Twenty-first Century* (Tokyo, United Nations University Press), 355–75; Gehring, T. and Buck, M. (2002), 'International and Transatlantic Environmental Governance', in Buck, M., Carius, A. and Kollmann, K. (eds), *International Environmental Policymaking* (Munich, Ökom Verlag), 21–43.

19   Downie and Levy (n. 18 above), 358–9.

In response to the critiques of UNEP several suggestions for reform or possible alternatives have been proposed. Some of these include:

1. Strengthening the UN general assembly by reconstituting one of its committees as an environmental committee.[20]
2. Creating a standing commission on the environment and development or an office of international environmental ombudsman.[21]
3. Reforming the Trusteeship Council into an Environmental Trusteeship Council.[22]
4. Incorporating environmental threats into the mandate of the Security Council or for the Security Council to convene special sessions on environmental insecurities.[23]
5. Creating a UN based environmental protection council with binding law enforcement capabilities.[24]
6. Creating a global ministerial environment forum (GMEF).[25]
7. Replacing UNEP with a World Environment Organization (WEO, also termed Global Environment Organization).[26]
8. Strengthening of international regimes.[27]
9. Clustering of multilateral environmental agreements (MEAs).[28]

---

20 See Elliott (n. 1 above), 103.

21 See for example, Schrijver, N. (1989), 'International Organization for Environmental Security', *Bulletin of Peace Proposals* 20:2, 115–22.

22 For example, Imber, M. (1994), *Environment, Security and UN Reform* (London, Macmillan), 106 referring to comments made by the UNCED Secretary-General Maurice Strong; Palmer, G. (1992), 'New Ways To Make International Environmental Law', *American Journal of International Law* 86:2, 259–83, 279; United Nations Secretary-General – Annan, K. (1997), *Renewing the United Nations: A Programme for Reform*, UN Doc. A/51/950 (New York, United Nations Secretariat), para. 85; see also further below.

23 See for example, Elliott 2004 (n. 1 above), 103.

24 See for example, Palmer 1992 (n. 22 above), 279 (referring to a New Zealand government proposal).

25 Governing Council of the United Nations Environment Programme, 'Global Ministerial Environment Forum', SS VII/I International Environmental Governance, UNEP/GC/21 (February 2002).

26 A variety of WEO models have been proposed ranging from a hierarchical and centralized body with considerable powers to more loose organizational structures such as the clustering of MEAs and their secretariats. Some proposals also suggest incorporating UNEP *into* a WEO structure. For useful summaries of the various proposals, see especially Biermann, F. and Bauer, S. (eds) (2005), *A World Environment Organization* (Aldershot, Ashgate), 9–10; Charnovitz, S. (2002), *A World Environment Organisation* (Tokyo, United Nations University Institute of Advanced Studies); Lodefalk, M. and Whalley, J. (2002), 'Reviewing Proposals for a World Environment Organisation', *The World Economy* 25:5, 601–17; Simonis, U. (2002), 'Advancing the Debate on a World Environment Organisation', *The Environmentalist* 22:1, 29–42.

27 See for example, Young, O. (ed.) (1997), *Global Governance: Drawing Insights from the Environmental Experience* (Cambridge, MA, MIT).

28 See for example, Moltke, K. von (2005),'Clustering International Environmental Agreements as an Alternative to a World Environment Organization', in Biermann, F. and

Collectively, these suggestions for reform raise two key issues for the future of global environmental governance, namely, the role of the UN and the desirability (or lack thereof) of centralization.

Notably, the majority of the above suggestions are variants of UN reform. This retention of current UN structures and desire for incremental rather than radical reform of the governance system appears to be supported by many governments. The Cartagena Declaration of 2002, for example, suggested that 'the process (of institutional reform) should be evolutionary in nature ... A prudent approach to institutional change is required, with preference given to making better use of existing structures'.[29] Some commentators also state a clear preference for utilizing and reforming existing institutions and structures rather than developing new international governance frameworks.[30] Not everyone agrees however, that the UN is a suitable actor. Falk, for example, argues that UN reform is impossible due to the inherent realist mindset and political preconditions operating within the organization.[31] He notes the geo-political closure (lack of consensus, problematic leadership and US dominance), the UN charter (too rigid) and the fundamentally hierarchical and patriarchal structure of the organization as key obstacles to any meaningful reform.[32]

In relation to the issue of centralization, two key themes have emerged. First, the perceived need for an authoritative environmental body with the capacity to control and deploy resources.[33] Second, the increased need for effective coordination

---

Bauer, S. (eds) (2005), *A World Environment Organization* (Aldershot, Ashgate), 175–204.

29   UNEP Governing Council, seventh special session (Cartagena Declaration) (Cartagena, Colombia, 15 February 2002; Decision SS.VII/1), available at <http://www.unep.org/IEG/docs/IEG_decisionSS_VII_1.doc>, Appendix, para. 8(b), accessed 1 December 2007.

30   See for example, Ayre, G. and Callway, R. (eds), (2005) *Governance for Sustainable Development: A Foundation for the Future* (London, Earthscan), 205 (arguing that 'the collective case for strengthening and reforming our current processes is a far stronger one than for either the development of new institutions or for states to act independently'); Gehring, T. and Oberthür, S. (2000), 'Was bringt eine Weltumweltorganisation? Kooperationstheoretische Anmerkungen zur institutionnellen Neuordnung der internationalen Umweltpolitik', *Zeitschrift für Internationale Beziehungen* 7:1, 185–211 arguing in favour of using resources we have through current structures; Gehring, T. and Oberthür, S. (2005), 'Reforming International Environmental Governance: An Institutional Perspective on Proposals for a World Environment Organization', in Biermann, F. and Bauer, S. (eds) (2005), *A World Environment Organization* (Aldershot, Ashgate), 205–35; Najam, A. (2005), 'Neither Necessary, nor Sufficient: Why Organizational Tinkering Will Not Improve Environmental Governance', in Biermann, F. and Bauer, S. (eds) (2005), *A World Environment Organization* (Aldershot, Ashgate), 235–56.

31   Falk, R. (1999), *Predatory Globalization: A Critique* (Cambridge, Polity Press), 111–12.

32   Ibid. 113–15.

33   The Commission on Global Governance (1995), *Our Global Neighbourhood* (Oxford, Oxford University Press), 4; see also various authors (n. 15 above) in support of a centralized WEO.

between various governance actors due to the growth, diversification and increasingly complex system of international environmental governance.[34]

The extent of centralization, and the desirability (or lack thereof), is usefully illustrated in the debate surrounding proposals for a WEO. Centralization has been noted as having several key advantages, for instance:

- Concentration of resources – which may increase the ability to develop and utilize compliance mechanisms, and give greater ability to impose sanctions for non-compliance.[35]
- Concentration of power – allowing a challenge to other powerful international actors such as the WTO or IMF to become possible.[36]
- Consolidation of information – regimes and MEAs centralized in one place – easing access and administrative burden, avoiding crossover and duplication.[37]
- Increased uniformity of international sustainability norms and principles.

The arguments for centralization (especially on certain issues, such as, for example, water or climate) to increase effectiveness and efficiency in administration are compelling.[38] Yet centralization may also be criticized as maintaining the problems commonly associated with hierarchical organizational structures, such as lack of accountability and transparency and skewed power dynamics in decision-making, agenda-setting and prioritization of issues.[39] Overall, it appears that suggestions for less centralized types of global governance structures tend to place more emphasis on, and allow more room for, the importance of coordination between various governance actors (including non-state actors and interest groups) and current governance structures (for example, clustering of MEAs). Rather than focusing on the possible gains that a centralized governmental organization may bring, arguments in favour of less centralized global governance structures note the importance of greater integration and dialogue between existing structures.[40]

Despite varying differences in focus (at times diametrically opposing), proponents of centralized and non-centralized reform proposals have in common their recognition for the importance of a change in ethics, be this effected through

---

34 See for example, Ayre and Callway 2005 (n. 30 above), 28; Elliott 2004 (n. 1 above), 102. These arguments are sometimes referred to as advocating *international pluralism*.

35 See for example Kanie, N. and Haas, P. (eds) (2004), *Emerging Forces in Environmental Governance* (Tokyo, United Nations University Press), 272; Report of the Commission on Global Governance (1995), *Our Global Neighbourhood* (Oxford, Oxford University Press), 4.

36 Kanie and Haas 2004 (n. 35 above), 272.

37 Ibid.

38 See for example, Ayre and Callway 2005 (n. 30 above).

39 For criticism of centralization see, for example, Gehring and Oberthür 2000 (n. 30 above); Gehring and Oberthür 2005 (n. 30 above); Moltke 2005 (n. 28 above); Newell, P. (2002), 'A World Environment Organisation: The Wrong Solution to the Wrong Problem', *The World Economy* 25, 659–71.

40 See for example, Ayre and Callway 2005 (n. 30 above), 28.

reform of existing institutional structures or the development of new institutions. The need for increased transparency and strengthening of the democratic framework; greater coordination across policy areas and institutions; and better enforcement of non-economic treaties such as MEAs are noted as key issues that must be addressed by any form of future global governance if sustainability is to become reality.[41]

Irrespective of the form(s) of governance adopted at a global level, be it reform of the UN, development of a WEO or the clustering of MEAs, for example, a central guiding principle must be north–south equity. In general, it appears fair to state that centralization has been viewed with disfavour in terms of achieving north–south equity as the power dynamics in centralized global governance structures usually operate to disadvantage countries from the global south.[42] As Gupta notes, often the issues that dominate the international agenda on global environmental governance are reflective of Western domestic agendas.[43] Furthermore, due to weaker bargaining power, the global south is frequently faced with major negotiating problems resulting in an inability to challenge such inequities in agenda setting.[44] Andrew Simms notes this as the potential conflict between differential prioritization of interests of north and south. He cites the example of a focus on climate change, driven by the north, in the face of a reality for many people in the global south for whom issues such as acute poverty, lack of healthcare and education may be higher priorities.[45] Indigenous communities have also frequently experienced their participation in international negotiations as frustrating, as their interests and rights have gained little government recognition or address.[46]

These considerations are a reminder that sustainability cannot be perceived independently from social and environmental justice which forms the basis of institutional governance structures. The redress of the current dichotomy between the global north and south has priority in any attempt at governance reform.

---

41   Ibid. (Arguing that we need 'greater vertical and horizontal integration of dialogue and decision-making across organisations'.)

42   See for example, Gupta, J. (2005), 'Global Environmental Governance: Challenges for the South from a Theoretical Perspective', in Biermann, F. and Bauer, S. (eds) (2005), *A World Environment Organization* (Aldershot, Ashgate), 57–86, arguing against a WEO on the premise that centralization will work against the interests of the global south.

43   Ibid. 65.

44   Gupta 2005 (n. 42 above), 78.

45   See for example, Barry, J. and Eckersley, R. (eds) (2005), *The State and the Global Ecological Crisis* (Cambridge, MA, MIT), 270; Simms, A. (2005), 'Economy: The Economic Problem of Sustainable Governance', in Ayre, G. and Callway, R. (eds), *Governance for Sustainable Development: A Foundation for the Future* (London, Earthscan), 103–25, 103.

46   See for example, Fogel, C. (2004), 'The Local, the Global, and the Kyoto Protocol', in Jasanoff, S. and Martello, M.L. (eds), *Earthly Politics: Local and Global in Environmental Governance* (Cambridge, MA, MIT), 103–25, 103: discussing indigenous peoples' relationship to climate change negotiation she argues: 'Indigenous peoples saw themselves as unrepresented in the discourse of either the Inter-governmental Panel on Climate Change (IPCC) or the official Kyoto protocol negotiations. Responding to this they demanded that governments acknowledge their existence and contributions, the specialized knowledges that they hold and their rights to participate in global climate change institutions that impact on their sovereign territories'.

Another key issue of more effective governance is enforcement and judicial review. The International Court of Justice (ICJ), the principal judicial organ of the United Nations[47] is underutilized in the pursuit of environmental governance.[48] It has the power to adjudicate on legal disputes referred to it and to provide advisory opinions to an organization of the United Nations if so requested. It must work within the framework of the general principles and purposes set out in the UN Charter and has equal footing with the other principal organs.[49] The other organs of the UN are careful not to involve the Court in issues that are prominently political.

The ICJ is not a general 'constitutional court' of the UN. Each organ of the United Nations is free to interpret the Charter as and when circumstances require. Therefore it does not function as an appellate court from the decisions of the General Assembly or the Security Council, nor does it function as an authority for judicial review to determine the 'constitutionality' of the actions or decisions of any other organ or subdivision of the United Nations.

Since the 1980s there have been many attempts to open access to the Court to a wider community or to create mandatory jurisdiction. And there have been repeated calls for an independent environmental court. The latest is the proposal of the Brazilian government for an 'International Environmental Court', modelled after the ICJ in The Hague, Netherlands. The court would seek to balance economic development with protecting the environment, punishing environmental crimes on a global level.[50] Similarly, the Australian Greens have proposed to 'establish an international environmental court and an environmental council at the UN, with similar decision-making powers to the Security Council, that deals with environmental issues of global significance'.[51]

A promising initiative of the UN was the aforementioned proposal to reconstitute the defunct UN Trusteeship Council. Based on earlier suggestions by Maurice Strong in 1988[52] and the Commission on Global Governance in 1995,[53] the idea was to establish 'a collective trusteeship for the integrity of the global environment and common areas'.[54] The proposal was referred to a Task Force chaired by the Executive

---

47   Charter of the United Nations (San Francisco, 26 June 1945), Article 92.

48   Hempel, L. (1996), *Environmental Governance: The Global Challenge* (New York, Island Press), 176.

49   Statute of the International Court of Justice 1945, Article 65, para. 1.

50   Presented at the Conference of the Americas for the Environment and Sustainable Development, September 2004, see <www.worldchanging.com/archives/001225.html>, accessed 1 December 2007; see also Postiglione, A. (2001), 'An International Court for the Environment', in Gleeson, B. and Low, N. (eds), *Governance for the Environment: Global Problems, Ethics and Democracy* (London, Palgrave), 211–20.

51   Item 16 of the Green Policy on Global Governance, available at <http://greens.org. au/election/policy.php?policy=40>, accessed 1 December 2007.

52   Strong, M.F. (1989), 'The United Nations in an Interdependent World', *International Affairs*, January, 11–21, 20.

53   Report of the Commission on Global Governance (n. 35 above), 251.

54   Report of the Secretary-General to the General Assembly, 'Renewing the United Nations: A Programme for Reform' (14 July 1997; UN Doc. A/51/950), para. 85; see also Chapter 5.

Director of UNEP, but not taken further.[55] Instead, the UNEP Governing Council referred the matter to the Open-ended Intergovernmental Group of Ministers on Environmental Governance from where it was referred to a consultation meeting of experts. These experts, in their wisdom,[56] concluded that 'it would be very difficult to undertake measures that would affect the main organs established by the United Nations Charter, like ECOSOC and the Trusteeship Council'.[57] The winding up of the Trusteeship Council in 2005[58] may have put an end to this initiative.

However, the case for global trusteeship or guardianship models remains strong. In the absence of any legal status for the global environment or common areas (oceans, atmosphere), there is no *erga omnes* status of global environmental responsibility. This vacuum has never been filled through treaty negotiations or administrative measures. Instead, all negotiated regimes are not only vague and patchy, but reinforce the states' sovereign right of exploitation. The assumed legitimacy and legality of using the Earth's resources has been questioned by many commentators,[59] but not by states. The only way out of the dilemma between territorial sovereignty and environmental responsibility is the recognition of the fundamental principle of sustainability. Only then would the right of use replaced by a right of *sustainable use*.[60]

When the former Prime Minister of New Zealand, Sir Geoffrey Palmer, suggested a 'new organ in the United Nations system ... empowered to take binding decisions on global environmental issues', he called this 'a conceptual leap forward in institutional terms'.[61] That was in 1989, this revolutionary year when

55   The report of the Task Force on Environment and Human Settlements to the General Assembly omitted any recommendation; Report of the Secretary-General to the General Assembly, 'Environment and Human Settlements' (6 October 1998; UN Doc. A/53/463).

56   Sand, P. (2006), 'Global Environmental Change and the Nation State', in Winter, G. (ed.), *Multilevel Governance and Global Environmental Change* (Cambridge, Cambridge University Press), 519–38, 530.

57   Estrada Oyuela, R. (2001), 'Expert Consultations on International Environmental Governance', Cambridge, 28–9 May, Chairman's summary (UNEP IEG Working Document), 1.

58   UN GA Res.60/1 of 16 September 2005 (World Summit Outcome).

59   For example, Gillespie, A. (1997), *International Environmental Law, Policy and Ethics* (Oxford, Clarendon Press), Chapter 2; see also Gillespie, A. (2002), 'International Environmental Law and Policy', in Bosselmann, K. and Grinlinton, D. *Environmental Law for a Sustainable Society* (Auckland, New Zealand Centre for Environmental Law), vol.1, 67–78, describing the sovereign right to exploit as 'radically unsuited to the demands of the twenty-first century' (67) and 'one of the largest contradictions of our time' (77).

60   Birnie, P. and Boyle, A. (2002), *International Law and the Environment*, 2nd edn (Oxford, Oxford University Press), 89: efforts to 'crystallize' a principle of sustainable utilization of natural resources from the general development of international environmental law (citing, for example, the concept of sustainable development) are bound to fail as long as the fundamentality of the principle of sustainability is overlooked.

61   Palmer, G. (1989), Verbatim Record of the 15th Meeting GA, 44th Session, 61 (UN Doc. A/44/PV.15); see also Palmer, G. (1995), *Environment: The International Challenge* (Wellington, Victoria University Press), 45–84.

even the G7 Summit thought that this 'new institution may be worth considering'.[62] Nearly 20 years later the 'conceptual leap forward' is more needed than ever. The current climate change negotiations, for example, are proof of the fact that states are compromising where conceptually no compromise is possible. The climate is either seen as destabilized or not. The Earth's ecological integrity is either threatened or not. These are issues directly addressed by the principle of sustainability as it provides the 'yardstick of permissible use of the environment'.[63]

The application of the principle of sustainability calls for decision-making powers of international bodies of trusteeship or guardianship. But this will only be possible if states themselves move towards environmental trusteeship and add ecological wisdom to their sovereignty.

The European Union

The example of the European Union (EU) demonstrates that states are, in fact, capable of reorganizing their sovereignty. For over 50 years the EU member states have noted the advantages to be gained by incrementally transferring sovereign rights to a supranational level. While these moves have been largely motivated by economic concerns, they have also improved the social and environmental well-being within member states and, as a consequence, at European level. Moreover, the EU is the world's only region where sustainable development combining economic, social and environmental policies is a declared constitutional objective. We can ask, therefore, to what extent the principle of sustainability has been implemented in the EU's governance.

One obvious advantage of the EU is that its member states coordinate their efforts, not through treaties and powerless institutions, but through legislation and centralized decision-making. In fact, the EU's administrative system represents a 'body [that] is established to act as a trustee of the regional interest'.[64] Such trustee function is, of course, far from genuine environmental trusteeship, but does demonstrate the potential of regional or even global environmental agencies acting on behalf of – then semi-autonomous – states. From this perspective, it is conceivable that states create supranational bodies that represent common areas, aspects of the global environment (for example, climate) or the global environment in its entirety. Such developments imply an 'enlightened' concept of territorial sovereignty and a fundamental acceptance of sustainability, but would not threaten state sovereignty. To this end, the EU provides a model of governance for sustainability.

---

62   Paragraph 8, G7: Summit of the Arch (1989), repr. in Starke, L. (ed.) (1990), *Signs of Hope: Working Together Towards Our Common Future* (Oxford, Oxford University Press).

63   Bothe, M. (2006), 'Whose Environment? Concepts of Commonality in International Environmental Law', in Winter, G. (ed.), *Multilevel Governance and Global Environmental Change* (Cambridge, Cambridge University Press), 539–58, 555.

64   Winter, G. (2006), 'Introduction', in Winter, G. (ed.), *Multilevel Governance and Global Environmental Change* (Cambridge, Cambridge University Press), 1–33, 24.

The EU was quick to reply to international calls for sustainability, responding to Rio by adding notions of sustainability into its constitutional framework.[65] The two first steps of this adoption of sustainability norms into the EU have been the fifth Environment Action Programme (EAP) and the Amsterdam Treaty on the European Union (TEU).

The fifth EAP 'explicitly takes up the definition of sustainable development proposed by the Brundtland Commission' and as such, is 'widely considered the firmest expression so far of this idea in the EU'.[66] The objective of the fifth EAP was to transform patterns of growth within the EU community in a manner that promotes sustainability.[67] Instrumentally, the EAP marked a departure from a 'command-and-control' approach in favour of 'shared responsibility between various actors – government, industry and the public – [which] is considered to be necessary to achieve progress towards sustainability'.[68] An example of this in practice was the Commission's creation of the European Consultative Forum on the Environment and Sustainable Development in 1997. 'This Forum is an independent advisory body with membership from NGOs, business and industry, consumers, farmers, local and regional authorities and academic communities'.[69] Despite novel intentions, however, the sixth EAP noted significant concern regarding the lack of willingness of member states to implement the fifth EAP.[70] In response to this concern the sixth EAP advocated 'a more inclusive approach including more specific targets and an increased use of market-based measures'.[71] In addition, the sixth EAP focused on better integration of environmental concerns into other policies.[72]

In an effort to strengthen implementation and compliance the Commission announced it would:

- increase pressure on member states by making implementation failures better and more widely known;
- encourage closer collaboration with the market; and
- ensure greater involvement of various stakeholders.[73]

---

65   Bosselmann, K. (2003), 'The Environmental Governance of the European Union: Institutional and Procedural Aspects of Sustainability', in Lilly, I. and Bosselmann, K. (eds), *Repositioning Europe: Perspectives from New Zealand*, NCRE Research Series No.2 (Christchurch, University of Canterbury) 9–30, 11.

66   Ibid.

67   See, 'Fifth European Community Environment Programme: Towards Sustainability', available at < http://europa.eu/scadplus/leg/en/lvb/l28062.htm>, accessed 1 December 2007.

68   Bosselmann 2003 (n. 65 above), 12.

69   Ibid. 19.

70   See, 'Sixth Environment Action Programme. Environment 2010: Our Future, Our Choice', available at <http://europa.eu.int/comm/environment/newprg>, accessed 1 December 2007.

71   Bosselmann, K. (2006), 'Missing the Point? The EU's Institutional and Procedural Approach to Sustainability', in Pallemaert, M. and Azmanova, A. (eds), *The European Union and Sustainable Development: Internal and External Dimensions* (Amsterdam, Kluwer), 105, 110.

72   Ibid. 111.

73   Bosselmann 2003 (n. 65 above), 12.

The second step of the EU's integration of sustainability norms is provided by the Amsterdam Treaty (TEU) which granted quasi-constitutional status to the idea of sustainability.[74] Of particular note are Article 6 and Article 3.[75] Article 6 proclaims the strengthening of the 'requirement of integrating environmental considerations into sectoral economic policies' by granting this 'the status of a basic organising principle of the Union'.[76] The revised Article 3 'clarifies that integration refers to all EU policies and activities laid out in Article 3. It explicitly links integration to the achievement of sustainable development'.[77] The formulation contained in the Amsterdam Treaty represents the cornerstone of the EU's current constitutional framework. The Draft European Constitution did not add anything of 'substance'.[78] In essence, the Draft Constitution failed to ensure the necessary repositioning of sustainability as the basic and fundamental organizing principle of all other policy. Although the integration of policy agendas in pursuit of sustainability has had a continually high profile, in practice the basic environmental paradigm has remained unchanged. Rather than emerging as a new guiding paradigm from which to reformulate policy, integration and institutional activity, the framework of sustainability has thus evolved as a mere extension of traditional environmental policies.[79]

The so-called Cardiff process, launched in 1998, concerns the integration of the environment into sectoral policies.[80] It is questionable, however, whether the integration process leads to (genuine) sustainable development or just to better coordination for three key reasons:

1. The Cardiff focus is on environmental integration whilst lacking essential defining elements that are necessary to achieve sustainability.
2. By centring around the Council, the Cardiff process leaves out other EU institutions and limits wider stakeholder involvement.
3. As the Cardiff focus is specifically directed at EU community policy it omits international, national and local dimensions.[81]

Despite several summits discussing the Cardiff process with a view to developing a comprehensive and integrated strategy for sustainability, it has remained unclear exactly how and to what degree the Cardiff process can be linked to the agenda of sustainability.[82] At the Copenhagen conference of the European Environment Agency (EEA) for instance, an attempt was made to link the Cardiff process and

---

74 Bosselmann 2006 (n. 71 above), 111.

75 Amsterdam Treaty 1997, available at <www.eurotreaties.com/amsterdamtext.html>, accessed 1 December 2007.

76 Bosselmann 2006 (n. 71 above), 111.

77 Ibid.

78 Ibid.

79 Ibid.

80 Ibid. 112.

81 Ibid.

82 Fergusson, M., Coffey, C., Wilkinson, D., Baldock, D. et al. (2001), *The Effectiveness of EU Council Integration Strategies and Options for Carrying Forward the Cardiff Process* (London, Institute for European Environmental Policy), xii, 66.

sixth EAP to the adoption of an overall sustainability strategy.[83] To ensure the increased coordination necessary for this to be successful, the EPA recommended a 'system of integrated monitoring and reporting' operating with 'headline indicators' that measure progress in the areas of structures, institutions and policies.[84] The fundamental criticism put forward by the EEA is that the EU has made little progress towards sustainability.[85] This criticism stems from EEA's vital recognition that sustainability 'will not come directly from environmental policies, but from socio-economic policies, guided by sustainability paradigms'.[86]

The failure to adequately accommodate the recognition stated above is illustrated clearly by the EU Strategy for Sustainable Development (SDS).[87] Once again, this EU initiative simply restates the ideal of sustainability as a key goal of the future, rather than establishing new principles to guide all policies *based on the paradigm of sustainability*. In part, this problem is compounded by lack of clear definition of sustainability itself. The SDS gives no definition for sustainable development or any indication of its content or guiding principles. As such, it does not resolve key issues of defining sustainability, in particular the exact relationship between economic, social and environmental spheres in the meaning of sustainability. As for the Cardiff process, the SDS appears to focus on increasing the effectiveness and efficiency of environmental policies but does not provide a paradigm shift towards sustainability.

In contrast to the Cardiff process and the SDS, the more recent White Paper by the Commission appears to create a possibility for a new approach to institutionalizing sustainability. Notably, the White Paper is in fact not directly concerned with the concept of sustainability, but is a reaction to the perceived dissatisfaction of EU citizens towards the Union's political institutions.[88] Despite the official focus of the White Paper being on generating 'good governance' in EU institutions however, some proposals are directly relevant to sustainability.[89] The five main principles of the White Paper are:

---

83   Bosselmann 2006 (n. 71 above), 113 referring to Jimenez-Beltran, D. (Executive Director, European Environment Agency), 'Making Sustainability Accountable: The Role and Feasibility of Indicators. From Gothenburg to Barcelona', Speech delivered in Brussels, 9 July 2001.

84   Ibid.

85   Ibid.

86   Jimenez-Beltran (n. 83 above).

87   Commission of the European Communities, 'A Sustainable Europe for a Better World: A European Strategy for Sustainable Development', (15 May 2001), available at <http://eur-lex.europa.eu/LexUriServ/site/en/com/2001/com2001_0264en01.pdf>, accessed 1 December 2007.

88   Commission of the European Communities, 'European Governance: A White Paper', (25 July 2001) 3, available at
   <http://eur-lex.europa.eu/LexUriServ/site/en/com/2001/com2001_0428en01.pdf>, accessed 1 December 2007.

89   Bosselmann 2006 (n. 71 above), 115.

1. openness;
2. participation;
3. accountability;
4. effectiveness; and
5. coherence.[90]

Incidentally, these principles are intended to reinforce two key principles of the EU, subsidiarity and proportionality.[91] According to Prue Taylor, the White Paper is an acknowledgement that the Union needs to 'phase out a top-down approach to policy- and regulation-making and to readdress the use of non-legislative instruments to implement policy'.[92] The White Paper's focus on decentralization, which is to be 'achieved not through delegation but by opening up the policy-making and policy-delivery processes to involve more people and organisations', is a reflection of this acknowledgement.[93] Thus, it should be noted that the White Paper attempts to step beyond a mere concern with improved policy, regulatory and enforcement mechanisms to address the more substantial need for a fundamental 'refocussing of institutions' in accordance with the framework of sustainability.[94] This is certainly an improvement on the Cardiff and SDS strategies exhibiting a greater complacency with 'tinkering' of policy and processes rather than presenting a paradigm shift.

Perhaps most importantly, the final part of the White Paper recognizes that governance is not only about processes but 'also about competence and power'.[95] In its final paragraphs the paper refers to building the Union upon a 'multi-level system of governance', stating that: 'In a multi-level system the real challenge is establishing clear rules for how competence is shared – not separated; only that non-exclusive vision can secure the best interests of all the Member states and all the Union's citizens'.[96] Thus, here we have a clear recognition of the need for a broadening of governance, in particular the heightened need for equitable communication, coordination and cooperation between different spheres, EU, national and local. Importantly, this push for a multi-level governance approach also appears consistent with the recognition that internal EU reform has more wide-ranging effects on global governance.[97] Given the indivisibility of the environment and the need for transnational action, this is an important recognition.

However, despite these positive trends, the White Paper is not without its problems. First, the goals set by the White Paper (for reforming Union institutions and processes) are arguably too modest and as such, likely to result in limited progressive

---

90   White Paper (n. 88 above).
91   Bosselmann 2006 (n. 71 above), 116.
92   Taylor, P. (2003), 'Reforming the Governance of the European Union: A Greater Voice and Expanded Planning Role for Local Government in EU Affairs?', in Lilly, I. and Bosselmann, K. (eds), *Repositioning Europe: Perspectives from New Zealand*, NCRE Research Series No.2 (Christchurch, University of Canterbury), 31–54, 33.
93   Ibid. 32–3.
94   Ibid. 33.
95   Ibid. 34.
96   White Paper (n. 88 above), 35.
97   Taylor 2003 (n. 92 above), 48.

innovation or fundamental change.[98] Second, although it is not explicitly stated, 'it is relatively clear that the EU Commission conceives the multi-level system of shared governance as primarily comprising the Commission and the Member states' thus continuing the exclusion of other stakeholders such as local government or civil society in governance processes.[99]

The importance of integrating local sectors into EU governance must not be underestimated. As argued by Taylor: 'Within Europe, the democratic ideal of the right of citizens to participate in the conduct of public affairs has long been recognised. It has also been acknowledged that it is at the local level that this right can be most directly exercised'.[100] The White Paper does acknowledge growing dissatisfaction amongst EU citizens with the current governance institutions, but as Taylor comments: 'The prospect of the EU becoming "broader and deeper" has compounded pre-existing feelings that governance is too centralised, that Europe is controlled by EU institutions at the expense of national and local institutions. This has resulted in feelings of alienation, mistrust and disinterest'.[101] The greater involvement of civil society and local forms of governance is crucial in mitigating this discontent and ensuring the successful implementation of sustainability norms into the broader EU institutional framework. On a conceptual level, for example, greater involvement of civil society in governance processes can help define the very concept of sustainability and help link this to institutional implementation.[102] On a more institutional level, widening the governance dialogue to ensure a diversification of stakeholder involvement (including the increased participation of civil society) can assist in advocating the need for a paradigm shift. Local government, for example, can play a key role in facilitating this shift.[103] Together, the better integration of all levels of government and the increased participation of civil society in governance processes can assist in ensuring that sustainability becomes the overarching paradigm from which governance design is to take place rather than sustainability being the mere goal that is the intended outcome of current governance structures.

Within the EU, there are numerous independent sustainability councils operating at national or regional level. More than 30 national councils cooperate in the network of European Environmental Advisory Councils (EEACs) to enrich the quality of policy advice at EU, national and regional levels. The EEACs' viewpoint on sustainability is expressed in the following statement:

> The basic principle of sustainability, EEACs believe, is that the natural environment has critical and unique values that can seldom be substituted by, or traded for, the economic or social products of civilisation. Sustainable development can be achieved only if the EU adopts a new concept of development, involving far-reaching modifications in patterns of both production and consumption. This new concept of development will acknowledge

98   Ibid. 34.
99   Ibid. 44.
100   Ibid. 37.
101   Ibid.
102   Bosselmann 2006 (n. 71 above), 124.
103   See generally, Taylor 2003 (n. 92 above).

economic needs and social aspirations, but accept protection of the environment and natural resources as fundamental.[104]

This and other statements of civil society groups express the voice of the principle of sustainability, a voice that the EU has not sufficiently listened to.

In conclusion, the EU is a model for substantial advances in structuring governance for sustainability, but equally for the continued failure of governments to capture its values and principles.

Actors for Change

The changes described so far can be promoted at international levelbut they usually originate within countries. The relationship between international and national governance structures is dialectic (both levels influence each other), but governments are likely to follow their own constituents and voters more than abstract international bureaucracies. In their analysis of how international environmental institutions may change their policy, Haas et al. found that domestic advocacy is still crucial:

> If there is one key variable accounting for policy change it is the degree of domestic environmentalist pressure in major industrialized democracies, not the decision-making rules of the relevant international institution. However, we do find that the institutions that have given rise to the most dramatic changes in collective decision-making are those that were able to apply constructive channels for such domestic pressure to reach governments.[105]

The 'constructive channels' are presumably those created by activists at home and accommodated at international institutional level. In their investigation into the domestic sphere, Haas et al. make out certain conditions for success. They label them the 'three Cs', namely concern, capacity and contract environment.[106] If a government has sufficient 'concern', it may provide administrative 'capacity' to eventually allow some form of a 'contract' between itself and civil society. Of these three conditions, governmental concern is vital.[107] Unless a government is under sufficient pressure, especially from media or recognized experts, it will not be concerned.

So, how to make governments concerned? From the point of view of ecological governance, the general acceptance of the sustainability idea among states shows that resonances do exist. There is, for example, common ground between weak sustainability approaches of governments and strong sustainability approaches of civil society. This common ground exists in two respects. First, there seems to be a commonly shared concern for a wider perspective of space and time: affluent societies should care for the poor (at home and abroad) and care for the future. Second, there

---

104   See Bosselmann 2006 (n. 71 above), 105–25, 125.

105   Haas, P., Keohane, R. and Levy, M. (eds) (1993), *Institutions for the Earth: Sources of Effective International Environmental Protection* (Cambridge, MA, MIT Press), 14.

106   Ibid. 11.

107   Ibid. 19.

seems to be acceptance that environmental, social and economic policies need to be (more) integrated.

Over the last ten years, most governments in industrialized countries have made some arrangements for more integrated policies. There are administrative arrangements within government, independent advisory bodies (of civil society) and national strategies for sustainable development.

The limits of such policy integration are drawn by economics. Economic growth and competitiveness continue to set boundaries to ethical-political implications of the principle of sustainability rather than the other way round. That sustainability sets boundaries to economic development is a truth that governments have not yet accepted. For this to happen the leadership of individual actors is required.

There is empirical evidence to suggest that there are certain mechanisms of 'momentum-building'. International environmental policies are formulated through conferences, reports, articles, etc. The reason why certain matters can be put onto the international agenda is the penetrating power of some actors. These can be 'lead' states, influential NGOs or persistent individuals such as Al Gore, for example. If the time is right and penetration strong enough, the agenda is set.[108]

According to political theorists, the business of agenda-setting involves lead actors and oppositional forces. The more recognized the actors appear, the more likely that opposition forms itself as a group in order to exercise a veto. This can lead to a 'veto bloc'.[109] In between acting and vetoing blocs is an uncommitted or undecided group typically referred to as a 'swing group'. The interplay of lead actors, veto blocs and swing groups can be observed in most processes of political decision-making.

One example is the political and legal success of the animal rights movement. Peter Singer's famous tractate *Animal Liberation* (1975) triggered it off. As James Jasper and Dorothy Nelkin observed in *The Animal Rights Crusade: The Growth of a Moral Protest*, 'Philosophers served as midwives of the animal rights movement in the late 1970s'. Thirty years later, Singer describes the 'generally favourable course of the philosophical debate about the moral status of animals'.[110] He refers to changed attitudes and practices in Europe and a generally better treatment of animals than 30 years ago. He hastens to add though, that more animals suffer today than ever and that 'popular views on that topic are still very far from adopting the basic idea that the interests of all beings should be given equal consideration irrespective of their species'.

---

108   Porter, G. and Brown, J.W. (1996), *Global Environmental Politics*, 2nd edn (Boulder, CO, Westview Press), 32; Hurrel, A. and Kingsbury, B. (1992), *The International Politics of the Environment* (New York, Oxford University Press), Chapter 1.

109   Porter and Brown 1996 (n. 108 above).

110   Singer, P., 'Animal Liberation at 30', *New York Times Review of Books*, Vol. 50, No. 8 (15 May 2003), 4, available at <http://www.nybooks.com/articles/16276>, accessed 1 December 2007.

Another example is Germany. As we have seen,[111] the German constitutional debate reflects both support and opposition to this basic idea. Eventually, a compromise was found in the form of Article 20a of the *Grundgesetz*.

Lead actors have been even more successful in Switzerland. A 1992 amendment to the Federal Constitution required the state to take into account the *Würde der Kreatur*.[112] This notion might be translated to 'dignity of creation',[113] except this resembles the German term *Schöpfung* (reflecting Christian terminology). The other official language, French, captures the idea of *Würde der Kreatur* much better: 'l'intégrité des organismes vivants'. This 'integrity of living organisms' would be very close to ecological integrity. However, there is ongoing dispute about the true meaning of *Würde der Kreatur*.

The main proponent of this constitutional move was Professor Peter Saladin, a constitutional lawyer, who had advocated ecological governance for a long time. According to him *Würde der Kreatur* has an essential core that may not be infringed upon and may not be set aside by a balancing process. In his 1994 report for the Swiss Environmental Protection Agency (EPA) he emphasized that *Würde des Menschen* and *Würde der Kreatur* do not point to something substantively different.[114] Both reflect intrinsic value and dignity. Not surprisingly, the EPA ordered a second opinion. It came from a Zurich University ethics group arguing that *Würde der Kreatur* is meant to be seen at a different level from human dignity. The group's 1997 report[115] called for the more narrow interpretation of dignity that protects individuals from degradation. Time will tell whether full recognition of ecological integrity remains on the agenda.

The Swiss discussion shows how a single advocate can set an agenda for constitutional change. The combination of sound argument, persistence and ceasing a momentum may be the key to success. The German and Swiss experiences also suggest at least the possibility for profound change. Governance for sustainability may gain traction, most likely through lead states such as these two, some Scandinavian countries, the Netherlands and New Zealand. The next

---

111   See Chapter 4.

112   The new Swiss constitution of 2000 incorporates the equivalent 1992 article as Article 120 ('gene technology in the nonhuman area'):

> 1. Persons and their environment shall be protected against abuse of gene technology.
> 2. The Confederation shall legislate on the use of the reproductive and genetic material of animals, plants, and other organisms. In doing so, it shall take into account the *dignity of creation* and the security of man, animal and environment, and shall protect the genetic multiplicity of animal and vegetal species.

113   A translation offered by the Swiss government.

114   Praetorius, I. and Saladin, P. (1994), *Die Würde der Kreatur*, Schriftenreihe Umwelt Nr. 260, BUWAL (ed.), 121; see also the standard commentary on the Swiss constitution in Saladin, P. and Schweizer, R. J. (1995), *Kommentar zur Bundesverfassung der Schweizerischen Eidgenossenschaft*, Article 24, novies Abs. 3 (Basle, Verlag Helbing & Lichtenhahn).

115   Published in English as Balzer, P., Rippe, K.P. and Schaber, P. (2000), 'Two Concepts of Dignity for Humans and Non-Human Organisms in the Context of Genetic Engineering', *Journal of Agricultural and Environmental Ethics* 13, 7–27.

step is then to envisage these states putting sustainability on the international agenda.

Behind these (and many other) signs of hope are ordinary people, legally called citizens. We will now see how an advanced understanding of citizenship can and arguably will create the critical mass for change.

Ecological Citizenship and Emerging Global Civil Society

For our context, citizenship can be defined as the 'membership, determined by factors such as place of birth, parentage or naturalisation, of a political community (generally a nation-state), in virtue of which one has legally defined rights and duties, significant identity and (on some accounts) moral responsibility to participate in public affairs'.[116] A citizen is thus 'a person, who is a member of a political community, owing allegiance to the community and being entitled to enjoy all its civil rights and protections'.[117]

Citizenship can also be described 'as both a set of practices (cultural, symbolic and economic) and a bundle of rights and duties' and it is stressed that it is neither a purely sociological nor a purely legal concept but is based on a relationship between the two[118] and expands to civil, social and political domains.

In addition, the term 'citizenship' is often used in a rather metaphorical sense to express the more general idea of membership in a community[119] and the quality of a person's conduct as such a member.[120]

In a formal legal sense, it describes a status conferred by the political community that constitutes the sovereign power and is legally ascribed to a certain group of individuals who by this means are bound together and distinct from other individuals. Thus, the status includes certain individuals and excludes others.

Substantially, citizenship confers rights and duties as instruments of the political community to generate internal security, stability and identity, create a sense of loyalty and, through taxation and military services, provide for the resources necessary for the survival and functioning of the community.[121] International Law has always respected the state's sovereign right to exclusively decide on questions of citizenship, i.e. to define under which conditions someone may or may not be a citizen.[122]

---

116   From the Glossary as used in Dower, N. (2003), *An Introduction to Global Citizenship* (Edinburgh, Edinburgh University Press), xi.

117   Garner, B. (ed.) (1999), *Black's Law Dictionary*. 7th edn (St Paul, MN, West Group), 237.

118   Isin, E. and Wood, P.K. (1999), *Citizenship and Identity* (London, Thousand Oaks and New Delhi, Sage Publications), 4.

119   Dower 2003 (n. 116 above), 36.

120   Garner 1999 (n. 117 above), 237.

121   Dunkerley, D., Hodgson, L., Konopacki, S., Spybey, T. and Thompson, A. (2002), *Changing Europe: Identities, Nations, Citizens* (London and New York, Routledge), 10–11.

122   Ottonelli, V. (2002), 'Immigration: What does Global Justice Require?', in Dower, N. and Williams, J. (eds), *Global Citizenship: A Critical Reader* (Edinburgh, Edinburgh

Traditionally, the legal basis of citizenship is either based on *ius sanguinis* or *ius soli*.[123] *Ius sanguinis* ('right or law of blood line'), signifies that someone 'inherits' their parents' citizenship irrespective of birthplace. It reflects an ethno-cultural understanding in which genealogy is the key to identification, resulting in a 'community of descent' characterized by its reluctance to absorb new members'.[124]

In contrast, *ius soli* ('right or law of soil'), is based on the principle of territoriality and automatically bestows citizenship of the country on a person born in that country. Identity as a citizen is seen from a political-territorial standpoint, as a product of environment rather than genes.[125]

Nations no longer strictly adhere to one concept but tend to mingle them, one as the general rule and the other as its exception. Today, the rules of gaining citizenship are more politically motivated.

In the following section, we will explore the dynamics and changing rules of citizenship. There is much to suggest that citizenship has lost its original function, i.e. to define people's rights and duties in their community. We should remind ourselves how crucially important this function remains. Without a mechanism that guarantees people the essential rights of living in their community, people and communities would lose all sense of security. The core idea of citizenship is to provide legal security and political identity.

This is the reason why most national constitutions acknowledge the fundamental status conveyed by citizenship and thus contain provisions preventing its denial.

Correspondingly, Article 15 of the Universal Declaration of Human Rights[126] states that a right to nationality is a fundamental human right.

Globalization and Citizenship

Globalization and migration have posed a fundamental challenge to identity, humanitarian law and human rights.[127] More people than ever before are faced

---

University Press), 231–43, 234 speaking of the principle of self-determination of political communities when granting membership; Dehousse, R. (1994), *Europe after Maastricht: An Ever Closer Union?* (Munich, Beck), 12.

123   See Heater, D. (1999), *What is Citizenship?* (Cambridge, Polity Press), 80.

124   Krieken, R. van (2000), 'Citizenship and Democracy in Germany: Implications for Understanding Globalisation', in Vandenberg, A. (ed.), *Citizenship and Democracy in a Global Era* (London, Macmillan Press), 131.

125   For example, France has traditionally practiced *ius soli* whereas Germany used to rely on *ius sanguinis*, or, as Heater puts it: 'On the French model citizens can be made; on the German, they must be born'. Germany now combines both models.

126   Adopted and proclaimed by General Assembly, Res. 217A (III) of 10 December 1948, whereafter the Assembly called upon all member countries to publicize the text of the Declaration and 'to cause it to be disseminated, displayed, read and expounded principally in schools and other educational institutions, without distinction based on the political status of countries or territories' (Universal Declaration of Human Rights) (UN Doc. A/810 (1948)), available at <http://www.un.org/Overview/rights.html>, accessed 1 December 2007.

127   See Dower 2003 (n. 116 above), 3; Falk, R. (2002), 'An Emergent Matrix of Citizenship: Complex, Uneven, and Fluid', in Dower, N. and Williams, J. (eds), *Global*

with their greatest fears, i.e. losing their political identity and sense of belonging. Ironically, the more open and globalized the world becomes, the greater the need for the protection of citizenship.

Citizenship is challenged by a host of political threats: international migration and mobility, increasingly multiracial populations, nationalism, intolerance, religious fundamentalism, loss of social and environmental security, democratic deficits, voter apathy and worldwide restructuring of welfare systems to accommodate demographic changes. In this situation it becomes once again clear that human rights and democracy are dependent on the quality of the citizens: 'The institutions of constitutional freedom are only worth as much as a population makes of them'.[128]

Citizenship, in turn, depends on its ability to respond to such changes. To start with, individuals can no longer simply turn to the state for the fulfilment of their basic aspiration. Citizenship has become 'less organically connected to an individual's search for security and meaning in life'.[129]

(1) Proliferation of identity   In a globalized world, states cannot rely on their citizens' national identity. Instead, we observe a proliferation of identities in one person, the multi-belonging citizen. We are (more or less) all part of a 'global culture'[130] and the internationalization of values. Both are facilitated through globally produced goods and global culture factories ('Hollywood') preaching the gospel of individualism, consumerism and hedonism.[131]

The counter-forces of economic globalization are disloyal identities and different worldviews. Global culture is made up of competing paradigms, i.e. the modern paradigm of economism versus the postmodern paradigm of ecologism. The ecological paradigm fits a world embracing diversity within unity. We encounter this new world on an almost daily basis. Most of us live in societies in which many people travel all over the world and encounter other people from all over the world. To some extent, it becomes impossible to ascertain how much foreign culture is 'absorbed' with each encounter. The foreign becomes the familiar, and the familiar becomes the foreign. As a result, national identity is being replaced by a multinational identity more akin to a global citizen.

Citizenship: A Critical Reader (Edinburgh, Edinburgh University Press); Hilson, C. (2000), 'Greening Citizenship: Boundaries of Membership and the Environment', *Journal of Environmental Law* 13, 335–48, 335.

128   Kymlicka, W. and Norman, W. (1994), 'Return of the Citizen: A Survey of Recent Work on Citizenship Theory', *Ethics* 104, 352–81, 353 quoting Habermas.

129   Falk 2002 (n. 127 above), 16.

130   Isin and Wood 1999 (n. 118 above), 155.

131   Castles, S. and Davidson, A. (2000), *Citizenship and Migration: Globalization and the Politics of Belonging* (London, Macmillan Press), 8. On the other hand, there are also indications that cultural diversity will prevail because of 'rapidity and multidirectionality of mobility and communication': Isin and Wood 1999 (n. 118 above), 127.

(2) Shifting loyalty Closely connected to identity is loyalty, as its emotional, intellectual or psychological prerequisite to form a bond of alliance with a set of values encapsulated in a system.

Loyalty is defined as the act of binding oneself to a course of action, a vow, cause or principle. Commitment and dedication are closely intertwined with loyalty, and related issues are communalism, devotion, faithfulness, nationalism, patriotism and steadfastness.[132]

Just as with identity, loyalty has become either weak or split, often detached from nation, state and territory and more conditional than it used to be, in the sense that it can be challenged more easily. Loyalty nowadays frequently expresses critical commitment rather than blind obedience.[133] Moreover, loyalty is not exhausted in its application to a specific nation or state, but refers to values rather than institutions and to institutions only to the extent to which they embody or protect such values. Loyalty can expand as well as shrink depending on the point of reference. It can apply to one's family in equal terms as to humanity as a whole, depending on the sphere or layer of moral obligation.

Loyalty is often accompanied by a feeling of moral obligation and responsibility as with loyalty comes the desire to protect and worship the ideals towards which one feels loyal. This can go as far as to elicit an altruistic readiness of self-sacrifice for a 'higher cause'. With respect to global problems such as poverty, war, ecological decline, social injustice and human rights violations, loyalty can be the ferment for a sense of global justice uniting humanity in an unprecedented manner.

(3) New citizenship laws Reflecting non-national or multinational identity and loyalty are those legal concepts that promote transferral of political powers. This is happening with all kinds of supra- or trans- and international organizations, such as the European Union,[134] the World Trade Organization, the World Bank and the United Nations, to name but a few. State sovereignty has become 'increasingly pooled'.[135] Other forms of political connections include economic free trade zones (such as APEC), military alliances (such as NATO), cultural and educational programmes (such as UNESCO), but also groupings such as the 'Western industrialized countries', 'the Arab countries', the 'Middle' and the 'Far East' – even the infamous 'Axis of Evil' or the 'Old Europe'.

In this sense, political and legal perceptions of identity and loyalty have become more closely connected to broader concepts.[136] They also demonstrate the potential

---

132 For definitions see <www.hyperdictionary.com/dictionary/loyalty>, accessed 1 December 2007.

133 For a summary see Waller, M. and Linklater, A. (2003), *Political Loyalty and the Nation-State* (London and New York, Routledge), 228–9.

134 Economides, K. et al. (eds) (2000), *Fundamental Values: A Volume of Essays to Commemorate the 75th Anniversary of the Founding of the Law School in Exeter* (Oxford, Hart Publishing), 116 speaks of a 'seepage of power'.

135 Hilson 2000 (n. 129 above), 336.

136 Kritsiotis, D. (2000), 'Imagining the International Community', *EJIL* 13, 961–92, 962 critically refers to the 'emerging sense of global community' and at 966–7 and 991–2 questions the term 'international community' altogether; Hilson 2000 (n. 127 above), 337

of multiple and multi-layered, extra-national forms of citizenship, with European integration being the most advanced model.[137] In combination with the principle of subsidiarity, regional arrangements such as the EU are conceived of as a new federal structure replacing the representative nation-state system.[138] Moreover, the individual has been given procedural rights such as the access to the European Court of Justice and the right to take legal action before national courts to enforce the realization of EU guidelines.[139] It is this novelty that is interpreted as an indicator of the emergence of a new layer of citizenship.[140] EU citizenship is sometimes considered the 'guinea pig' or even precursor to global citizenship.

Considering the darker side of European integration ('Fortress Europe'), some say that European citizenship 'may in fact prove to be a hindrance to global citizenship' as many of the policies of EU citizenship run counter to the idea of cosmopolitan citizenship.[141] So far, Union citizenship has not succeeded in uniting different nationalities. It relies on traditional notions (such as nationality itself), and while a European identity may be emerging, it comes with a sense of exclusion: 'A specific type of identity is being fostered. It excludes long-term legally resident third country immigrants or citizens who do not meet certain economic conditions'.[142] Some warn there is a real danger that the population in the EU will be divided in citizens and an underclass of 'foreigners'.[143]

Nevertheless, the European experience represents a helpful reference for global citizenship.[144] Whereas different ideologies form the basis of these types of citizenship, namely with regard to the degree of institutionalization and inclusion or exclusion respectively, some features resemble each other, such as the idea that transnational citizenship could be complementing national citizenship.

---

considers the environment, the poor, women rather than Britain or Europe, as examples of identification.

137   Carter, A. (2001), *The Political Theory of Global Citizenship* (London and New York, Routledge), 119, 139.

138   Davidson, A. (2000), 'Democracy, Class and Citizenship', in Vandenberg, A. (ed.), *Citizenship and Democracy in a Global Era* (Basingstoke, Macmillan), 117–8, 120.

139   Economides et al. 2000 (n. 134 above), 124–5.

140   Castles and Davidson 2000 (n. 131 above), 176; Dehousse 1994 (n. 122 above), 147 observes a significant departure from the traditional link between nationality and citizenship of the nation-state, a loosening of the metaphysical ties between persons and a state, and forming a symptom of cosmopolitization of citizenship.

141   Steenbergen, B. van (1994), 'Towards a Global Ecological Citizen', in Steenbergen, B. van (ed.), *The Condition of Citizenship* (London, Sage Publications), 148; Carter 2001 (n. 137 above), 141, 235 reasons that regional bodies may obstruct as much as support international goals.

142   Dunkerley et al. 2002 (n. 121 above), 19.

143   Dehousse 1994 (n. 122 above), 146.

144   Ibid. 148 conceding that European citizenship may be useful as a laboratory for a modern active procedural concept of proto-cosmopolitan citizenship.

Global and Ecological Citizenship

The strongest justification for global citizenship is the globalized world we live in, more precisely, the global nature of problems we are facing. None of these can be solved by states and their citizens alone. States, like citizens, are almost forced into the (eco-)logic of global citizenship. At the very least, traditional citizenship needs to transform to a new model of multilayer citizenship.

There is little revolutionary about transforming citizenship. Even political liberalism has acknowledged this – either in practical terms ('free movement of capital, goods and services') or in theoretical terms observing the 'changing face of citizenship'.[145] Bart van Steenbergen promotes the 'global ecological citizen',[146] and Ralf Dahrendorf formulates the enormous challenge for a truly global concept of citizenship:

> In contrast to universal human rights, there is much less consensus as to the worth of these values, and they are not only centred on humanity itself but also on other living species or the Earth in general. The same complications occur when duties are concerned which a citizen not only owes to her home country but to the world community, to the earth, to humankind.[147]

If the expansion of citizenship is to tackle root causes of global problems, a theory of ecological global citizenship is needed. Some proponents of such a theory should be introduced here:

(1) Nigel Dower According to Dower, global citizenship consists of three components: a normative, an existential and an aspirational one.[148]

The normative aspect is based on the conviction that as global citizens, we all share certain duties, and that we all have a moral status of being worthy of moral respect.[149]

In its existential meaning, the idea acknowledges that we are all part of a global community, be it an institutional or quasi-political one.

The trend towards global integration has linked economic markets and thus political decision-making, a symbol of which is the annual economic summit of the seven leading industrial countries ('G7'). Moreover, there is an expanding consensus around the world that certain adjustments of the global pattern of consumption have to be made to save the human species from extinction.[150]

---

145   Dahrendorf, R. (1994), 'The Changing Quality of Citizenship', in Steenbergen, B. van (ed.), *The Condition of Citizenship* (London, Sage Publications), 13.

146   Steenbergen 1994 (n. 141 above).

147   Dahrendorf 1994 (n. 145 above), 18.

148   Dower 2003 (n. 116 above), 140.

149   Ibid., 6, 147, 149.

150   Ibid. 54, 124, 140 et seq.

The aspirational side is based on a long tradition of thought and feeling about the ultimate unity of human experience.[151] It rests upon the other two in striving for a world in which basic values become fully realizable through strengthening of the community and its institutions and building a legal framework.

Dower understands global citizenship as comprising all human beings in virtue of their immanent human rights and duties. However, only some actively use their status as global citizens through institutions of global civil society. Through exercise of responsibility or assertion of universal rights, and certain attitudes towards human beings, varying degrees of 'active' global citizenry are possible.[152] One of the main goals of global citizenship is the further enhancement of its institutions.[153]

His approach thus embodies both passive and active membership and the institutional implications of the concept. It therefore is a philosophical attitude-related idea as well as a political mission.

(2) Richard Falk[154] identifies four different levels 'to conceive the extension of citizenship beyond the traditional boundaries of nation and state'.[155]

One is the aspirational and normative feature as well as a tendency towards global and especially economic integration – just as in Dower's concept. Moreover, Falk calls for the adoption of 'politics of impossibility based on attitudes of necessity to avoid human extinction in the face of destructive consumptive patterns' and a politics of mobilization.[156]

Based on these four levels, he distinguishes five, partly overlapping, but also contrary concepts.[157] First, there is the global reformer who believes in a better way of organizing the political life of the planet and accepts some political centralization to overcome current problems.[158] Then there is the elite global businessman who is a product of the impact on identity of globalization of economic forces in a world which is unified around a denationalized elite. This elite shares interests and experiences, but lacks civic responsibility – and is thus the very opposite of a global citizen.[159] Third, there is the manager of the world order who has emerged in reaction to environmental problems and is modelled after the Brundtland report.[160] Fourth, the politically conscious rationalist[161] is emerging. Lastly, there is the transnational

---

151   Ibid. 7; Falk, R. (1994), 'The Making of Global Citizenship', in Steenbergen, B. van (ed.), *The Condition of Citizenship* (London, Sage Publications), 131.

152   Dower 2003 (n. 116 above), 6–7.

153   Dower, N. (2002), 'Global Citizenship: Yes or No?', in Dower, N. and Williams, J. (eds), *Global Citizenship: A Critical Reader* (Edinburgh, Edinburgh University Press), 124, 140.

154   Falk 1994 (n. 151 above), 127–140.

155   Ibid. 131.

156   Ibid. 131–2.

157   Ibid. 139.

158   Ibid. 132–4; Steenbergen 1994 (n. 141 above), 149 calls this type of citizen a *potential* ecological citizen.

159   Falk 1994 (n. 151 above), 134–5.

160   Ibid. 135–6.

161   Ibid. 136–9.

activist who originates from the grassroots phenomenon of civil society and tries to stimulate a process of globalization from below.[162]

(3) Ralf Dahrendorf[163] criticizes recent 'confused thinking' to fight grievances by entitlements and thus to let lawyers or judges do the work politicians ought to do in the first place.[164]

He defines the ecological question of citizenship as one of human habitat and concludes that citizenship can only be complete if seen in its global dimension. The main problem he acknowledges is how to combine environmental protection and sustainable development with certain entitlements of citizenship which are linked to economic expansion, i.e. how it is possible to curb unfettered expansion without violating other entitlements of citizenship.[165]

(4) Bart van Steenbergen[166] differentiates between three approaches towards ecological citizenship: one of increasing inclusion, meaning the challenge of the idea that only existing human beings can be citizens, an approach followed by the animal rights movements; one concentrating on human responsibility for nature, a policy which is often promoted by green parties emphasizing the existence not only of social but also ecological responsibility; and a third one stressing the global dimension of ecological citizenship.[167]

In addition, Steenbergen offers two types of global environmental citizens: the Earth citizen, who is aware of the Earth as a living organism (Gaia) and her origin from Earth, which leads to the concept of care instead of control and humans as participants instead of subjugators – and the world citizen who perceives of the environment as a matter of 'big science' and the planet an object of global management which requires large scale organization and government.[168] Thus, Steenbergen emphasizes the attitude-centred side of citizenship.

(5) Aldo Leopold,[169] in his landmark A Sand County Almanac, analyses what it means to live in harmony with the land and with one another. Leopold's understanding of conservation is that of a 'state of harmony between men and land'. By 'land', he means all of the things Earth as one organism has to offer. 'Harmony with land is like harmony with a friend; you cannot cherish his right hand and chop off his left'. His

---

162 Steenbergen, B. van (1994), 'The Condition of Citizenship: An Introduction', Steenbergen, B. van (ed.), *The Condition of Citizenship* (London, Sage Publications), 7 referring to Falk 1994 (n. 151 above), 138–9.

163 Dahrendorf 1994 (n. 145 above), 10–19.

164 Ibid. 18.

165 Ibid.

166 Steenbergen 1994 (n. 141 above), 141–52.

167 Ibid. 143–51.

168 Ibid.

169 See <www.aldoleopold.org/> and <www.naturenet.com/alnc/aldo.html>, accessed 1 December 2007; see also Leopold, A. (1949), *A Sand County Almanac* (Oxford, Oxford University Press).

holistic view was based on the logic that if the land mechanism as a whole is good, then every part is good irrespective of humankind's knowledge of its function.[170]

He therefore considered a land ethic to change the role of *Homo sapiens* from conqueror of the land-community to plain member and citizen of it. This land ethic implies respect for his fellow-members, and also respect for the community as such.

Notwithstanding important philosophical differences between these authors, they agree not only on the validity of ecological citizenship, but also on many prerequisites for it: the normative idea of a community of humankind, the assumption of a community of life, the existential recognition that the future of the human species depends on the functioning of its surrounding ecosystems, and an aspiration of an increasing sense of responsibility that will lead to action on behalf of the environment.

If it is true that transforming attitudes and ideas precede the transformation of institutions, we can go a step further and describe some characteristics of ecological citizenship.

Ecological citizenship can be conceived as an extension of individual citizenship. We can imagine ourselves as both citizens of a social community and of an ecological community. And both forms of community can be experienced locally, nationally, regionally or globally. Thus, ecological citizenship is not global as opposed to national, but relevant at any community level that we are part of.

The defining aspect of ecological citizenship is the recognition of non-human beings as 'fellow citizens'. The notion of non-human citizens is purely metaphoric, but helpful to acknowledge a fiduciary relationship between citizenship and non-human entities. Ecological citizenship adopts a guardianship responsibility for entities not represented in the political decision-making process.

The responsibility of the 'ecologically aware citizen' has been described as follows:

> The ecologically aware citizen takes responsibility for the place where … she lives, understands the importance of making collective decisions regarding the commons, seeks the common good, identifies with bioregions and ecosystems rather than obsolete nation-states or transnational corporations, considers the wider impact of … her actions, is committed to … community building, observes the flow of power … and acts according to his or her conviction.[171]

While this statement describes ecological awareness, it does not define ecological citizenship. As a legal and political concept, it needs clear definition.

The fiduciary aspect of citizenship requires, first of all, wide-reaching procedural and participatory rights. The Earth Charter describes them in Principle 13:

> Strengthen democratic institutions at all levels, and provide transparency and accountability in governance, inclusive participation in decision making, and access to justice.

---

170   Leopold 1949 (n. 169 above), 145–6.
171   Presented in Isin and Wood 1999 (n. 118 above), 117 quoting Tomashow, M. (1995), *Ecological Identity: Becoming a Reflective Environmentalist* (Cambridge MA, MIT), 139.

a. Uphold the right of everyone to receive clear and timely information on environmental matters and all development plans and activities which are likely to affect them or in which they have an interest.
b. Support local, regional and global civil society, and promote the meaningful participation of all interested individuals and organizations in decision making.
c. Protect the rights to freedom of opinion, expression, peaceful assembly, association, and dissent.
d. Institute effective and efficient access to administrative and independent judicial procedures, including remedies and redress for environmental harm and the threat of such harm.
e. Eliminate corruption in all public and private institutions.
f. Strengthen local communities, enabling them to care for their environments, and assign environmental responsibilities to the levels of government where they can be carried out most effectively.

A further requirement of the fiduciary aspect is the recognition of duties. As much as there are rights to access and use the environment, there are limitations of these rights. Limitations to rights are not unusual, but an inherent part of human rights operating in a social context. The novelty is the acceptance that human rights operate in an ecological context as well. The limiting factor is the preservation of the Earth's ecological integrity.

The Earth Charter expresses these limitations in various respects:

Principle 1 (a)

Recognize that all beings are interdependent and every form of life has value regardless of its worth to human beings

...

Principle 2 (a)

Accept that with the right to own, manage, and use natural resources comes the duty to prevent environmental harm and to protect the rights of people ...

Principle 6 (a)

Place the burden of proof on those who argue that a proposed activity will not cause significant harm, and make the responsible parties liable for environmental harm.

These and other principles of the Earth Charter put self-restrictions on citizenship rights. In order to exercise our fiduciary responsibility, the right to the use of natural resources, for example, is restricted by respect for the community of life and the duty to prevent harm.

Consequently, the concept of ecological citizenship is embedded in the Earth Charter. Having described the individual components of global ecological governance, we can now describe the collective component, i.e. global civil society.

Emerging Global Civil Society

When citizens 'move' they constitute a citizens' movement. Citizens' movements and civil society groups are the driving force for social change. The perspective of governments and states defines them in more reserved terms: non-governmental organizations (NGOs) making 'important contributions in many fields, both nationally and internationally'[172]

According to April Carter, the concept of civil society evolved in the 1980s when dissident intellectuals in Eastern Europe created a social platform for resistance and political change.[173] If civil society was responsible for the most radical system change in recent times, it is perceivable that it is capable of more. It is certainly recognized today as the main driving force towards global governance.

In international environmental law, the pressures of civil advocacy are particularly strong. This legal field would not even exist were it not for the worldwide environmental movement putting global environmental degradation on the agenda of state conferences. The existence and practice of the UN Environmental Programme (UNEP), for example, is largely shaped through the input of environmental groups.[174]

Global civil society builds upon the autonomy of civil society bodies within their own nation states and links them within a transnational realm independent of all nation states.[175] It represents the whole network of international relationships and organizations which underlie society outside the sphere of established political institutions.[176]

While globalization has curtailed nation-states' capacity to regulate key areas, it has also opened up new spaces to be filled by other actors. States are no longer just tolerating civil society groups, they depend on them to make up for lost political power. In this respect, globalization thus cuts both ways. Originating from these grassroot movements, sometimes very powerful NGOs[177] have succeeded in setting the international agenda.

However, despite the positive influence global civil society can have on global governance, it lacks democratic legitimacy and accountability in its classic sense. Ultimately, while its policies might impact on citizens' lives, citizens in turn do not

---

172   Report of the Commission on Global Governance (n. 35 above), 32–33.

173   Carter 2001 (n. 137 above), 79.

174   Barcena, A. (1997), 'Global, Environmental Citizenship', UNEP 25, *Our Planet* No. 8, January, shows how UNEP's Global Environmental Citizenship Programme has developed strategic alliances with parliamentarians, consumers, local authorities, educators, religious groups, media and other key civil society groups.

175   Carter 2001 (n. 137 above), 80.

176   Attfield, R. (2002), 'Global Citizenship and the Global Environment', in Dower, N. and Williams, J. (eds), *Global Citizenship: A Critical Reader* (Edinburgh, Edinburgh University Press), 191–200, 197.

177   Apart from NGOs, civil society consists of various other groups such as trade unions, business associations, religious bodies, academic institutions, student organizations, ethnic lobbies, community groups, and so forth. Similarly defined in Dower, N. and Williams, J. (eds), *Global Citizenship: A Critical Reader* (Edinburgh, Edinburgh University Press), xxi.

have any democratic means to control civil society.[178] Moreover, civil society has mainly been dominated by Western and northern states and does therefore carry the imminent danger of endorsing moral concepts entertained by the more privileged parts of global society.[179] Its global expansion could thus bring about the exportation of predominantly Western values, Westernization and cultural imperialism.[180]

Civil society, in its present form, cannot substitute a representative system of governance, but could be legitimized by an emerging global citizenship. The more the concept of global citizenship is being associated with global civil society, the stronger its mandate becomes. As Robin Attfield points out, legitimizing and monitoring institutions of global governance could be the main effect of global citizenship.[181] It is certainly possible to imagine global governance without states.

Taking a more state-centred perspective, even the Commission on Global Governance is convinced that global governance without civil society is impossible. Its definition of global governance clearly involves civil society:

> Governance is the sum of the many ways individuals and institutions, public and private, manage their common affairs. It is a continuing process through which conflicting or diverse interests may be accommodated and co-operative actions may be taken. It includes formal institutions and regimes empowered to enforce compliance, as well as informal arrangements that people and institutions either have agreed to or perceive to be in their interest. … At the global level, governance has been viewed primarily as intergovernmental relationships, but it must be understood as also involving nongovernmental organizations (NGOs), citizens' movements, multinational corporations, and the global capital. Interacting with these are the global mass media of dramatically enlarged influence.[182]

Conclusion

The Earth Charter clearly assumes the leadership role of global citizens and global civil society.[183] They lead where states and state-related institutions must follow. The Earth Charter's ultimate, yet reconciliatory message to states is expressed in 'The Way Forward': 'Life often involves tensions between important values. This can mean difficult choices. However, we must find ways to harmonize diversity with unity, the exercise of freedom with the common good, short-term objectives and long-term goals'.

These sentiments leave room for modest reforms as much as for radical change. Either way, the Earth Charter represents civil society's profound commitment to

---

178   Isin and Wood 1999 (n. 118 above), 117.

179   Ibid. 121–2.

180   Ibid., pointing to the fact that, so far, globalization has meant Westernization.

181   Attfield 2002 (n. 176 above), 200.

182   Commission on Global Governance (n. 35 above), 2–3. The inclusion of 'multinational corporations, and the global capital' is, of course, the crux of most political moves towards global governance.

183   Bosselmann, K. (2007), 'The Earth Charter and Global Environmental Governance', in Gupta, K., Jankowska, M. and Bosselmann, K. (eds), *Global Environment: Problems and Policies* (New Delhi, Atlantic Publishers), 55.

sustainability. If states are not able and willing to support this commitment and exercise leadership, humanity is heading for disaster. Humanity would either be destroyed by the powers of militant states or by the powers of nature. A new governance for sustainability is the only choice we have.

Bibliography

Aarnio, A. (1990), 'Taking Rules Seriously', *ARSP* 42, 180–92.

Abel, W. (1976), *Die Wüstungen des ausgehenden Mittelalters*, 3rd edn (Stuttgart, Fischer).

Agius, E. (1998), 'Towards a Relational Theory of Intergenerational Ethics', in Busuttil, S. et al. (eds), *Our Responsibilities to Future Generations* (Malta, Foundation for International Studies).

Agudo, C. (2003), *Environment and Human Rights Report*, Committee on the Environment, Agriculture and Local and Regional Affairs, (16 April 2003; Doc. 9791) <http://assembly.coe.int/Documents/WorkingDocs/doc03/EDOC9791. htm>, accessed 1 December 2007.

Allott, P. (1990), *Eunomia: New Order for a New World* (New York, Oxford University Press).

Almond, B. (1995), 'Rights and Justice in the Environmental Debate', in Cooper, D. and Palmer, J. (eds), *Just Environments – Intergenerational, International and Inter-Species Issue*s (London, Routledge).

Annan, K. (1997), *Renewing the United Nations: A Programme for Reform*, UN Doc. A/51/950 (New York, United Nations Secretariat).

Archer, J.H. and Jarman, C. (1992), 'Sovereign Rights and Responsibilities: Applying Public Trust Principles to the Management of EEZ Space and Resources', *Journal of Ocean and Shoreline Management* 17, 251–64.

Arrow, K., Bolin, B., Costanza R. et al. (1995), 'Economic Growth, Carrying Capacity, and the Environment', *Science* 268, 520–21.

Attfield, R. (2002), 'Global Citizenship and the Global Environment', in Dower, N. and Williams, J. (eds), *Global Citizenship: A Critical Reader* (Edinburgh, Edinburgh University Press).

Austin, J. (1995), *The Province of Jurisprudence Determined*, ed. Rumble, W. (Cambridge, Cambridge University Press).

Ayre, G. and Callway, R (eds) (2005) *Governance for Sustainable Development: A Foundation for the Future* (London, Earthscan).

Bachmann, G. (2006), 'Warum blieb der Kollaps im neuzeitlichen Deutschland aus?', *GAIA* 15:4, 260–63.

Bahro, R. (1972), *Die Alternative* (Frankfurt and Cologne, European Publishing House).

Bahro, R. (1994), *Avoiding Social and Ecological Disaster: The Politics of World Transformation: An Inquiry into the Foundations of Spiritual and Ecological Politics*, tr. David Clarke (Bath, Gateway).

Balzer, P., Rippe, K.P. and Schaber, P. (2000), 'Two Concepts of Dignity for Humans and Non-Human Organisms in the Context of Genetic Engineering', *Journal of Agricultural and Environmental Ethics* 13, 7–27.

Barberis, J. (1988), 'Le Regime juridique international des eaux souterainnes', *Annuaire Francais de Droit International* 33, 129–62.

Barcena, A. (1997), 'Global, Environmental Citizenship', UNEP 25, *Our Planet* No. 8.

Barry, B. (1995), *Justice as Impartiality* (Oxford, Clarendon Press).

Barry, J. (2001), 'Greening Liberal Democracy: Practice, Theory and Political Economy', in Barry, J. and Wissenburg, M. (eds), *Sustaining Liberal Democracy: Ecological Challenges and Opportunities* (Basingstoke, Palgrave).

Barry, J. and Eckersley, R. (eds) (2005), *The State and the Global Ecological Crisis* (Cambridge, MA, MIT).

Bartelmus, P. (1994), *Environment, Growth and Development: The Concepts and Strategies of Sustainability* (Florence, KY, Routledge).

Baxter, B.H. (2000), 'Ecological Justice and Justice as Impartiality', *Environmental Politics* 9, 43–64.

Baxter, B.H. (2004), *A Theory of Ecological Justice* (London, Routledge).

Beckerman, W. (1974), *In Defence of Economic Growth* (London, Jonathan Cape).

Beckerman, W. (1994), 'Sustainable Development: Is it a Useful Concept?', *Environmental Values* 3, 191–209.

Beckerman, W. (1995), *Small is Stupid: Blowing the Whistle on the Greens* (London, Duckworth).

Beckerman, W. (1995), ' "How Would You Like Your 'Sustainability', Sir? Weak or Strong?" A Reply to My Critics', *Environmental Values* 5, 169.

Beesley, J.A. (1973), 'The Canadian Approach to International Environmental Law', *Canadian Yearbook of International Law* 11, 3–12.

Begon, M., Harper, J. and Townsend, C. (1996), *Ecology: Individuals, Populations and Communities*, 3rd edn (London, Blackwell Science).

Bell, D. (2002), 'How Can Political Liberals Be Environmentalists?', *Political Studies* 50, 703–24.

Berg, W. (1997), 'Typologie von Staatsbeschreibungen', in Burmeister, J. (ed.), *Verfassungsstaatlichkeit. Festschrift für Klaus Stern* (Munich, C.F. Beck).

Bericht der gemeinsamen Verfassungskommission, BT-Drucksache 12/6000, 661.

Bericht der Sachverständigenkommission 'Staatszielbestimmungen/ Gesetzgebungsaufträge', Rdnr. 144.

Bernhardt, R. (1976), 'Ungeschriebenes Völkerrecht', *Zeitschrift fur auslandisches offentliches Recht und Volkerrecht* 36, 50–76.

Beyerlin, U. (1996), 'The Concept of Sustainable Development', in Wolfrum, R. (ed.), *Enforcing Environmental Standards: Economic Mechanisms as Viable Means?* (Berlin, Springer).

Beyerlin, U. and Marauhn, T. (1997), *Law-Making and Law-Enforcement in International Environmental Law after the 1992 Rio Conference* (Berlin, Erich Schmidt Verlag).

Beyerlin, U. (2006), 'Bridging the North-South Divide in International Environmental Law', *Zeitschrift für auslandisches und offentliches Recht* 66, 263–75.

Beyerlin, U. (2007), 'Different Types of Norms in International Environmental Law: Policies, Principles and Rules', in Bodansky, D., Brunnée, J. and Hey, E. (eds), *Oxford Handbook of International Environmental Law* (Oxford, Oxford University Press).

Biermann, F. and Bauer, S. (eds) (2005), *A World Environment Organization* (Aldershot, Ashgate).

Birnie, P. and Boyle, A. (2002), *International law and the Environment*, 2nd edn (Oxford, Oxford University Press).

Boer, B. (1995), 'Implementation of International Sustainability at a National Level', in Ginther, K. et al. (eds), *Sustainable Development and Good Governance* (Dordrecht and Boston, Kluwer Academic Publishers).

Borggreve, B. (1888), *Die Forstabschätzung* (Berlin), 253.

Bosselmann, K. (1985), 'Wendezeit im Umweltrecht', *Kritische Justiz*, S. 356 f., 345.

Bosselmann, K. (1986), 'Eigene Rechte für die Natur?', *Kritische Justiz*, 1–22.

Bosselmann, K. (1992), *Im Namen der Natur* (Munich, Scherz).

Bosselmann, K. (1993), 'Introduction' in Stone, C. *Umwelt vor Gericht* (German tr. of *Should Trees have Standing?*), 2nd edn (Munich, Trickster Verlag).

Bosselmann, K. (1995), *When Two Worlds Collide: Society and Ecology* (Auckland, RSVP).

Bosselmann, K. (1995), 'Nichtanthropozentrische Erweiterung des Umweltverwaltungsrechts?', in Nida-Rümelin, J. and Pfordten, D. (eds), *Ökologische Ethik and Rechtstheorie* (Baden-Baden, Nomos).

Bosselmann, K. (1998), *Ökologische Grundrechte* (Baden-Baden, Nomos).

Bosselmann, K. (1999), 'Justice and the Environment: Building Blocks for a Theory on Ecological Justice', in Bosselmann, K. and Richardson, B. (eds), *Environmental Justice and Market Mechanisms: Key Challenges for Environmental Law and Policy* (London, Kluwer Law International).

Bosselmann, K. (2000), 'Un Approcio Ecologico Ai Diritti Umani', in Greco, M. (ed.), *Diritti Umani E Ambiente (Human Rights and the Environment)* (Fiersole, Edizioni Cultura della Pace).

Bosselmann, K. (2001), 'Human Rights and the Environment: Redefining Fundamental Principles?', in Gleeson, B. and Low, N. (eds), *Governance for the Environment* (London, Palgrave).

Bosselmann, K. (2001), 'Human Rights and the Environment: The Search for Common Ground', *Revista de Direito Ambiental*, 23 July–September, 12–28.

Bosselmann, K. and Schröter, M. (2001), *Umwelt und Gerechtigkeit* (Baden-Baden, Nomos).

Bosselmann, K. (2002), 'A Legal Framework for Sustainable Development', in Bosselmann, K. and Grinlinton, D. (eds) (2002), *Environmental Law for a Sustainable Society* (Auckland, New Zealand Centre for Environmental Law), vol. 1.

Bosselmann, K. (2002), 'The Concept of Sustainable Development', in Bosselmann, K. and Grinlinton, D. (eds) (2002), *Environmental Law for a Sustainable Society* (Auckland, New Zealand Centre for Environmental Law), vol. 1.

Bosselmann, K. and Grinlinton, D. (eds) (2002), *Environmental Law for a Sustainable Society* (Auckland, New Zealand Centre for Environmental Law), vol. 1.

Bosselmann, K. (2003), 'The Environmental Governance of the European Union: Institutional and Procedural Aspects of Sustainability', in Lilly, I. and Bosselmann,

K. (eds), *Repositioning Europe: Perspectives from New Zealand* (Christchurch, NCRE), vol. 2.

Bosselmann, K. (2004), 'Environmental Governance: A New Approach to Territorial Sovereignty', in Goldstein, R.J. (ed.), *Environmental Ethics and Law* (Aldershot, Ashgate).

Bosselmann, K. and Taylor, P. (2005), 'The Significance of the Earth Charter in International Law', in Blaze Corcoran, P. et al. (eds), *Toward a Sustainable World: The Earth Charter in Action* (The Hague, Kluwer International).

Bosselmann, K. (2006), 'Ecological Justice and Law', in Richardson, B. and Wood, S. (eds), *Environmental Law for Sustainability: A Critical Reader* (Oxford, Hart Publishers).

Bosselmann, K. (2006), 'Missing the Point? The EU's Institutional and Procedural Approach to Sustainability', in Pallemaert, M. and Azmanova, S. (eds), *The European Union and Sustainable Development: Internal and External Dimensions* (Amsterdam, Kluwer).

Bosselmann, K. (2007), 'Strong and Weak Sustainable Development: Making the Difference in the Design of Law', 13 *South African Journal of Environmental Law and Policy*, 14–23.

Bosselmann, K. (2007), 'Why New Zealand Needs a National Sustainable Development Strategy', in Parliamentary Commissioner for the Environment, *Sustainability Review 2007: New Zealand's Progress Toward Sustainable Development*, Background Paper, Wellington.

Bosselmann, K. (2007), 'The Earth Charter and Global Environmental Governance', in Gupta, K., Jankowska, M. and Bosselmann, K. (eds), *Global Environment: Problems and Policies* (New Dehli, Atlantic Publishers).

Bosselmann, K. (2008), 'The Environmental Jurisprudence of International Tribunals: Making Sustainability Count', in Kotze, L. and Robinson, N. (eds), *Compliance and Enforcement. Toward More Effective Implementation of Environmental Law* (Cambridge, Cambridge University Press).

Bosselmann, K. (2008), 'The Way Forward: Governance for Ecological Integrity', in Westra, L., Bosselmann, K. and Westra, R. (eds), *Reconciling Human Existence and Ecological Integrity* (London, Earthscan).

Bothe, M. (2006), 'Whose Environment? Concepts of Commonality in International Environmental Law', in Winter, G. (ed.), *Multilevel Governance of Global Environmental Change* (Cambridge, Cambridge University Press).

Boyle, A. (1996), 'The Rio Convention on Biological Diversity', in Bowman, M. and Redgwell, C. (eds), *International Law and the Conservation of Biological Diversity* (London, Kluwer Law International).

Brooks, R., Jones, R. and Virginia, R. (2002), *Law and Ecology: The Rise of the Ecosystem Regime* (Aldershot, Ashgate).

Brown, D. (2008), 'The Case for Understanding Inadequate National Responses to Climate Change and Human Rights Violations', in Westra, L., Bosselmann, K. and Westra, R. (eds), *Reconciling Human Existence and Ecological Integrity* (London, Earthscan).

Brown, L. (1981), *Building a Sustainable Society* (New York, W.W. Norton).

Brown, P.G. (2001), *The Commonwealth of Life: A Treatise on Stewardship Economics* (Montréal, Black Rose Books).

Brown Weiss, E. (1989), *In Fairness to Future Generations* (New York, Transnational).

Brown Weiss, E. (1989), *In Fairness to Future Generations: International Law, Common Patrimony and Intergenerational Equity* (United Nations University Press).

Brown Weiss, E. (1990), 'Intergenerational Justice and International Law', in Busuttil, S., Agius, E., Inglott, P.S. and Macelli, T. (eds), *Our Responsibilities to Future Generations* (Malta, Foundation for International Studies).

Brown Weiss, E. (1993), 'International Environmental law: Contemporary Issues and the Emergence of a New World Order', *Georgetown Law Journal* 81, 675–710.

Brown Weiss, E. (1995), 'Environmental Equity: The Imperative for the Twenty-First Century', in Lang, W. (ed.), *Sustainable Development and International Law* (Dordrecht, Martinus Nijhoff).

Brunee J. (1989), 'Common Interest – Echoes from the Empty Shell?', *ZaoRV* 49:4, 791–808.

Bundesamt für Umwelt, Wald und Landschaft (eds) (1995), *Die Würde der Kreatur* (Gutachten).

Burmeister, H. (ed.) (1994), *Wege zum ökologischen Rechtsstaat* (Taunusstein, Eberhard Blottner).

Bush, M. (2000), *Ecology of a Changing Planet*, 2nd edn (Upper Saddle River, NJ, Prentice Hall).

Caldwell, L. (1984), *International Environmental Policy: Emergence and Dimensions* (Durham, NC, Duke University Press).

Caldwell, L.K. (1963), 'Environment: A New Focus for Public Policy?', *Public Administration Review* 23, 132–9.

Callicott, J.B. (1990), 'The Case Against Moral Pluralism', *Environmental Ethics* 12, 99–115.

Calliess, C. (2001), *Rechtsstaat und Umweltstaat* (Tubingen, Mohr Soebeck).

Camilleri, J. and Falk, J. (1992), *The End of Sovereignty?: The Politics of a Shrinking and Fragmenting World* (Aldershot, Edward Elgar).

Carlowitz, H.C. von (1713), *Sylvicultura oeconomica. Anweisung zur wilden Baum-Zucht* (Leipzig; repr. Freiberg , TU Bergakademie Freiberg und Akademische Buchhandlung, 2000).

Carter, A. (2001), *The Political Theory of Global Citizenship* (London and New York, Routledge).

Carty, A. (1986), *The Decay of International Law?* (Manchester and Dover, NH, Manchester University Press).

Castles, S. and Davidson, A. (2000), *Citizenship and Migration: Globalization and the Politics of Belonging* (London, Macmillan Press).

Charnovitz, S. (2002), *A World Environment Organisation* (Tokyo, United Nations University Institute of Advanced Studies).

Cheng, B. (1953), *General Principles of Law as Applied by International Courts and Tribunals* (London, Stevens).

Chimni, B.S. (1998), 'The Principle of Permanent Sovereignty Over Natural Resources: Toward a Radical Interpretation', *Indian Journal of International Law* 38, 208–17.

Christian, S. (2007), in Working Group of the Parties to the Aarhus Convention, *Compilation of Responses to the Draft Strategic Plan*.

Cleveland, H. (1993), 'The Global Commons: A Global Commons Trusteeship Commission Is Needed to Guide our Use of the Oceans, Antarctica, the Atmosphere, and Outer Space', *The Futurist*, May–June, 9–13.

Collins, L. (2007), 'Are We There Yet? The Right to Environment in International and European Law', *McGill International Journal of Sustainable Development Law & Policy* 3, 119–153.

Commission of the European Communities (2001), 'A Sustainable Europe for a Better World: A European Strategy for Sustainable Development', 15 May 2001.

Commission of the European Communities (2001), 'European Governance: A White Paper', 25 July 2001.

Commission on Global Governance (1995), *Our Global Neighbourhood* (Oxford, Oxford University Press).

Cooper, D. (1995), 'Other Species and Moral Reason', in Cooper, D. and Palmer, J. (eds), *Just Environments – Intergenerational, International and Inter-Species Issues* (London, Routledge).

Cordonier Segger, M.C. and Khalfan, A. (2004), *Sustainable Development Law: Principles, Practices and Prospects* (Oxford, Oxford University Press).

Cotta, H. (1817), *Anweisung zum Waldbau*, 2nd edn (Dresden, Arnold).

Coyle, S. and Morrow, K. (2004), *Philosophical Foundations of Environmental Law* (Oxford, Hart Publishing).

Curran, S. (2004), 'Sustainable Development v Sustainable Management: The Interface between the Local Government Act and the Resource Management Act', *New Zealand Journal of Environmental Law* 8, 267–94.

D'Amato, A. (1990), 'Do We Owe a Duty to Future Generations to Preserve the Global Environment?', *American Journal of International Law* 84, 190.

D'Amato, A. and Chopra, S.K. (1991), 'Whales: Their Emerging Right to Life', *American Journal of International Law* 85, 21–62.

Dahrendorf, R. (1994), in Steenbergen, B. van (ed.), *The Condition of Citizenship* (London, Sage Publications).

Daly, H. (1991), 'Elements of Environmental Macroeconomics', in Constanza, D. (ed.), *Ecological Economics: The Science and Management of Sustainability* (New York, Columbia University Press).

Davidson, A. (2000), in Vandenberg, A. (ed.), *Citizenship and Democracy in a Global Era* (Basingstoke, Macmillan).

Dehousse, R. (1994), *Europe after Maastricht: An Ever Closer Union?* (Munich, Beck).

Diamond, J. (2005), *Collapse: How Societies Choose to Fail or Succeed* (New York, Viking Books).

Dobson, A. (1996), 'Environmental Sustainability: An Analysis and Typology', *Environmental Politics* 5, 401–28.

Donnelly, J. (2007), *International Human Rights*, 3rd edn (Boulder, CO, Westview Press).

Douzinas, C. and Warrington, R. (1994), *Justice Miscarried* (New York and London, Harvester Wheatsheaf).

Dower, N. (2002), 'Global Citizenship: Yes or No?', in Dower, N. and Williams, J. (eds), *Global Citizenship: A Critical Reader* (Edinburgh, Edinburgh University Press).

Dower, N. (2003), *An Introduction to Global Citizenship* (Edinburgh, Edinburgh University Press).

Downie, D.L. and Levy, M.A. (2000), 'The United Nations Environment Program at a Turning Point', in Chasek, P. (ed.), *The Global Environment in the Twenty-First Century* (Tokyo, United Nations University Press).

Dumandt, A. (1988), *Der Fall Roms – Die Auflösung des römischen Reiches im Urteil der Nachwelt* (Munich, Beck), 695.

Dunkerley, D., Hodgson, L., Konopacki, S., Spybey, T. and Thompson, A. (2002), *Changing Europe: Identities, Nations, Citizens* (London and New York, Routledge).

Dworkin, R. (1977), *Taking Rights Seriously* (Cambridge, MA, Harvard University Press).

Earth Charter Commission (2002), 'Earth Charter: Values and Principles for a Sustainable Future, The Earth Charter Initiative', <www.earthcharter.org> (Home Page), accessed 1 December 2007.

Eblinghaus, H. and Stickler, A. (1996), *Nachhaltigkeit und Macht. Zur Kritik von Sustainable Development* (Frankfurt, IKO).

Eckersley, R. (2001), 'Green, Justice, the State and Democracy', <www.arbld. unimelb.edu.au/envjust/papers/allpapers/eckersley/home.htm>., accessed 1 December 2007.

Eckersley, R. (2004), *The Green State: Rethinking Democracy and Sovereignty* (Cambridge, MA, MIT Press).

Economides, K. et al. (eds) (2000), *Fundamental Values: A Volume of Essays to Commemorate the 75th Anniversary of the Founding of the Law School in Exeter* (Oxford, Hart Publishing).

Elliot, R. (ed.) (1996), *Environmental Ethics* (Oxford, Oxford University Press).

Elliott, L. (2004), *The Global Politics of the Environment*, 2nd edn (New York, Palgrave Macmillan).

Ellis, J. and Wood, S. (2006), in Richardson, B. and Wood, D. (eds), *Environmental Law for Sustainability* (Oxford, Hart).

Emerson, J. 'What does Economic Rationality Do?', <http://www.idiocentrism. com/economic%20rationality.htm>, accessed 1 December 2007.

Epiney, A. and Scheyli, M. (1998), *Strukturprinzipien des Umweltvolkerrechts* (Bern, Stampfli).

Epiney, A. and Scheyli, M. (2000), *Umweltvolkerrecht* (Bern, Stampfli).

Estrada Oyuela, R. (2001), *Expert Consultations on International Environmental Governance*, UNEP IEG Working Document, 28–9 May 2001, Chairman's Summary (Cambridge).

Evelyn, J. (1664), *Sylva, or a Discourse of Forest-Trees and the Propagation of Timber in His Majesty's Dominions* (repr. 1776, London, Jo. Martyn and Ja. Allestry).

Experts Group on Environmental Law of the World Commission on Environment and Development (1988), *Environmental Protection and Sustainable Development, Legal Principles and Recommendations* (Dordrecht, Martinus Nijhoff Publishers).

FAO Code of Conduct on the Distribution and Use of Pesticides (November 2002), http://www.fao.org/ag/AGP/AGPP/Pesticid/Code/PM_Code.htm, accessed 1 December 2007.

Falk, R. (1989), *Revitalizing International Law* (Ames, IA, Iowa State University Press).

Falk, R. (1994), 'The Making of Global Citizenship', in Steenbergen, B. van (ed.), *The Condition of Citizenship* (London, Sage Publications).

Falk, R. (1999), *Predatory Globalization: A Critique* (Cambridge, Polity Press).

Falk, R. (2002), 'An Emergent Matrix of Citizenship: Complex, Uneven, and Fluid', in Dower, N. and Williams, J. (eds), *Global Citizenship: A Critical Reader* (Edinburgh, Edinburgh University Press).

Fastenrath, U. (1993), 'Relative Normativity in International Law', *EJIL* 4:1, 305–40.

Fergusson, M., Coffey, C., Wilkinson, D. and Baldock, D. (2001), *The Effectiveness of EU Council Integration Strategies and Options for Carrying Forward the 'Cardiff' Process* (London: Institute for European Environmental Policy).

Fifth European Community Environment Programme: Towards Sustainability (1992–2000), <http://europa.eu/scadplus/leg/en/lvb/l28062.htm>, accessed 1 December 2007.

Fogel, C. (2004), 'The Local, the Global, and the Kyoto Protocol', in Jasanoff, S. and Martello, M.L. (eds), *Earthly Politics: Local and Global in Environmental Governance* (Cambridge, MA, MIT).

Fondacaro, M. (2000), 'Toward an Ecological Jurisprudence Rooted in Concepts of Justice and Empirical Research', *UMKC Law Review* 69, 179–96.

Fong, G. (1993), 'Governance and Stewardship of the Living Resources: The Work of the South Pacific Forum Fisheries Agency', in Van Dyke, J.M., Zaelke, D. and Hewison, G. (eds), *Freedom for the Seas in the 21st Century: Ocean Governance and Environmental Harmony* (Washington, DC, Island Press), 131–41.

Foundation on Economic Trends (2001), Draft Treaty to Share the Genetic Commons, (Washington, DC, FOET; repr. in *South Letter* 2:37).

G7: Summit of the Arch (1989), repr. in Starke, L. (ed.) (1990), *Signs of Hope: Working Together Towards Our Common Future* (Oxford, Oxford University Press).

Garner, B. (ed.) (1999), *Black's Law Dictionary*, 7th edn (St Paul, MN, West Group).

Gebel, T. (1998), 'Der Treuhandgedanke und die Bewahrung der biologischen Vielfalt: Einschränkung der territorialen Souveränität durch treuhänderische Verwaltung von lebenden Umwelt-Ressourcen', Ph.D. diss. (Sinzheim, Pro Universitate Verlag).

Gehring, T. and Buck, M. (2002), 'International and Transatlantic Environmental Governance', in Buck, M., Carius, A. and Kollmann, K. (eds), *International Environmental Policymaking* (Munich, Okom).

Gehring, T. and Oberthur, S. (2000), 'Was bringt eine Weltumweltorganisation? Kooperationstheoretische Anmerkungen zur institutionnellen Neuordnung der internationalen Umweltpolitik', *Zeitschrift fur Internationale Beziehungen* 7:1, 185–211.

Gehring, T. and Oberthur, S. (2005), 'Reforming International Environmental governance: An Institutional Perspective on Proposals for a World Environment Organization', in Biermann, F. and Bauer, S. (eds) (2005), *A World Environment Organization* (Aldershot, Ashgate).

Giagnocavo, C. and Goldstein, H. (1989–90), 'Law Reform or World Re-form', *McGill Law Journal* 35, 345.

Gibson, N. (1990), 'The Right to a Clean Environment', *Saskatchewan Law Review* 54, 5–7.

Gillespie, A. (1997), *International Environmental Law, Policy and Ethics* (Oxford, Clarendon Press).

Gillespie, A. (2002), 'International Environmental Law and Policy', in Bosselmann, K. and Grinlinton, D. *Environmental Law for a Sustainable Society* (Auckland, New Zealand Centre for Environmental Law), vol.1.

Gillroy, J.M. (2006), 'Adjudication Norms, Dispute Settlement Regimes and International Tribunals: The Status of "Environmental Sustainability" in International Jurisprudence', *Stanford Journal of International Law* 42, 1–52.

Goethe, J.W. von (1801), *Wilhelm Meisters Lehrjahre*, 'Lehrbrief' in Buch 7, Kapitel 9.

Goldsmith, E. (1992), *The Way: An Ecological Worldview* (London, Rider).

Goldsmith, E., Allen, R., Allaby, M., Davoll, J. and Lawrence, S. (1972), *A Blueprint for Survival* (London, Tom Stacey).

Governing Council of the United Nations Environment Programme (2002), 'Global Ministerial Environment Forum', SS VII/I International Environmental Governance, UNEP/GC/21 February 2002.

Gowdy, J. (1999), *Coevolutionary Economics: The Economy, Society and the Environment* (Boston, Kluwer).

Green Policy on Global Governance (Australian) <http://greens.org.au/election/policy.php?policy=40>, accessed 1 December 2007.

Grinlinton, D. (2002), in Bosselmann, K. and Grinlinton, D. (eds), *Environmental Law for a Sustainable Society* (Auckland, New Zealand Centre for Environmental Law), vol. 1.

Grober, U. (2002), 'Tiefe Wurzeln: Eine Kleine Begriffsgeschichte von "sustainable development" – Nachhaltigkeit', *Natur und Kultur* 3:1, 116–28.

Grober, U. (2007), *Deep Roots – A Conceptual History of 'Sustainable Development' (Nachhaltigkeit)* (Berlin, Wissenschaftszentrum Berlin für Sozialforschung).

Grubb, M., Koch, M., Munson, A., Sullivan, F. and Thomson, K. (1993), *The Earth Summit Agreements: A Guide and Assessment* (London, Earthscan).

Grundy, K. (1995), 'Sustainable Management: A Sustainable Ethic?', Paper to the Third Annual Conference of the Resource Management Law Association, Wellington, New Zealand.

Gunn, A. and McCallig, C. (1997), 'Environmental and Social Justice in an Urban Setting – Sustainable Management and the New Zealand Resource Management Act 1991', Paper for the Conference 'Environmental Justice', University of Melbourne, 1–3 October.

Gupta, J. (2005), 'Global Environmental Governance: Challenges for the South from a Theoretical Perspective', in Biermann, F. and Bauer, S. (eds) (2005), *A World Environment Organization* (Aldershot, Ashgate).

Gupta, K, Jankowska, M. and Bosselmann, K. (eds) (2007), *Global Environment: Problems and Policies* (New Dehli, Atlantic Publishers).

Haas, P., Keohane, R. and Levy, M. (eds) (1993), *Institutions for the Earth: Sources of Effective International Environmental Protection* (Cambridge, MA, and London, MIT).

Habermas, J. (1991), *Erlauterungen zur Diskursethik* (Frankfurt, Suhrkamp).

Habermas, J. (1996), *Between Facts and Norms: Contributions to a Discourse Theory of Law and Democracy*, tr. Rehg, W. (Cambridge, Polity Press).

Habermas, J. (2001), *Die Zukunft der menschlichen Natur* (Frankfurt, Suhrkamp).

Hancock, J. (2003), *Environmental Human Rights* (Aldershot, Ashgate).

Handl, G. (1975), 'Territorial Sovereignty and the Problem of Transnational Sovereignty', *American Journal of International Law* 69, 50–76.

Handl, G. (1986), 'National Uses of Transboundary Air Resources', *Natural Resources Journal* 26, 405.

Handl, G. (1991), 'Environmental Security and Global Change: The Challenge to International Law', in Lang, W., Neuhold, H. and Zemanek, K. (eds), *Environmental Protection and International Law* (London, Graham & Trotman/ Martinus Nijhoff).

Hardie, W. (1980), *Aristotle's Ethical Theory* (Oxford, Oxford University Press).

Hardin, G. (1968), 'The Tragedy of the Commons', *Science* 162, 1243–8.

Hartig, G.L. (1795), *Anweisung zur Taxation der Forste oder zur Bestimmung des Holzertrags der Wälder* (Gießen).

Hayword, T. (1997), 'Interspecies Solidarity: Care Operated Upon by Justice', in Hayword T. and O'Neill, T. (eds), *Justice, Property and Environment* (Aldershot, Ashgate).

Heater, D. (1999), *What is Citizenship?* (Cambridge, Polity Press).

Heller, A. (1987), *Beyond Justice* (Oxford, Basil Blackwell).

Hempel, L. (1996), *Environmental Governance: The Global Challenge* (New York, Island Press).

Herder, J.G. (1784–91), *Ideen zur Philosophie der Geschichte der Menschheit*, ed. Martin Bollacher (Frankfurt, Deutscher Klassiker, 1985–2000).

Herlihy, D. (1998), *Der Schwarze Tod und die Verwandlung Europas* (Berlin, Wagenbach).

Hildreth, R.G. (1993), 'The Public Trust Doctrine and Coastal and Ocean Resources Management', *Journal of Environmental Law and Litigation* 8, 221–36.

Hilson, C. (2000), 'Greening Citizenship: Boundaries of Membership and the Environment', *Journal of Environmental Law* 13, 335–48.

Hohmann, H. (1994), *Precautionary Legal Duties and Principles of Modern International and Environmental Law* (London and Boston, Graham & Trotman).

Hughes, J.D. (2001), *An Environmental History of the World* (London, Routledge).

Hunter, D.B. (1988), 'An Ecological Perspective of Property: A Call for Juridical Protection of the Public's Interest in Environmentally Critical Resources', *Harvard Environmental Law Review* 12, 311–83.

Hurrel, A. and Kingsbury, B. (1992), *The International Politics of the Environment* (New York, Oxford University Press).

Huxley, T.H (1882) *Science and Culture and Other Essays* (New York, D. Appleton and Company).

Imber, M. (1994), *Environment, Security and UN Reform* (London, Macmillan).

International Court of Justice (1996), Legality of the Threat or Use of Nuclear Weapons, ICJ Advisory Opinion, 8 July.

International Union for Conservation of Nature and Natural Resources (IUCN) (ed.) (1980), *World Conservation Strategy: Living Resource Conservation for Sustainable Development* (Morges, Switzerland, IUCN).

International Union for the Conservation of Nature and Natural Resources (IUCN), World Wide Fund for Nature (WWF) and United Nations Environment Programme (UNEP) (1991), *Caring for the Earth: A Strategy for Sustainable Living* (Gland, Switzerland, IUCN).

Isin, E. and Wood, P.K. (1999), *Citizenship and Identity* (London, Thousand Oaks, Ca, and New Delhi, Sage Publications).

Jasper, J.M. and Nelkin, D. (1991), *The Animal Rights Crusade: The Growth of a Moral Protest* (New York, The Free Press).

Jenks, C.W. (1969), *New World of Law?* (London, Longmans Green).

Jimenez-Beltran, D. (2001), 'Making Sustainability Accountable: The Role and Feasibility of Indicators. From Gothenburg to Barcelona', Speech delivered in Brussels, 9 July.

Johnson, A. (1995), 'Barriers to Fair Treatment of Nonhuman Life', in Cooper, D. and Palmer, J. (eds), *Just Environments – Intergenerational, International and Inter-Species Issues* (London, Routledge).

Jonas, H. (1984), *The Imperative of Responsibility: In Search of an Ethics for the Technological Age*, tr. Jonas, H. and Herr, D. (Chicago, University of Chicago Press).

Kanie, N. and Haas, P. (eds) (2004), *Emerging Forces in Environmental Governance* (Tokyo, United Nations University Press).

Kingwell, M. (1995), *A Civil Tongue: Justice, Dialogue and the Politics of Pluralism* (University Park, PA, Pennsylvania State University Press).

Kirgis, F.L. (1972), 'Technological Changes to the Shared Environment', *American Journal for International Law* 66, 290–320.

Kiss, A. (1982), 'La Notion de patrimoine commun de l'humanité', *Recueil des Cours à l'Académie de Droit International* II, 99–256.

Kiss, A. (1983), 'The International Protection of the Environment', in MacDonald, R. and Johnston, D. (eds), *The Structure and Process of International Law* (The Hague, Martinus Nijhoff).

Kiss, A. (1985), 'The Common Heritage of Mankind: Utopia or Reality?' *International Journal* 40, 423–41.

Kiss, A. (1989), *Droit international de l'environnement* (Paris, Pedone).

Kiss, A. (1989), 'Nouvelles tendances en droit international de l'environment', *German Yearbook of International Law* 32, 241–63.

Kiss, A. (1999), 'International Trade and the Common Concern of Humankind', in Bosselmann, K. and Richardson, B. (eds), *Environmental Justice and Market Mechanisms: Key Challenges for Environmental Law and Policy* (London, Kluwer Law International).

Kiss, A. and Shelton, D. (2004), *International Environmental Law*, 3rd edn (New York, Transnational).

Kleinig, J. (1991), *Valuing Life* (Princeton, Princeton University Press).

Kloepfer, M. (ed.) (1994), *Umweltstaat als Zukunft. Studien zum Umweltstaat* (Bonn, Economica).

Kloepfer, M. and Mast, E. (1995), *Das Umweltrecht des Auslandes* (Berlin, Duncker & Humblot).

Korowicz, M. (1961/I), 'Some Present Aspects of Sovereignty in International Law', *Recueil des Cours à l'Académie de Droit International* 102, 1–120.

Kramer, L. (2004), 'The Genesis of EC Environmental Principles', in Macroy, R. (ed.), *Principles of European Environmental Law* (Groningen, Europa Law Publishing).

Krebs, A. (2000), 'Das teleologische Argument in der Naturethik', in Ott, K. and Gorke, M. (eds), *Spektrum der Umweltethik* (Marburg, Metropolis).

Krieken, R. van (2000), 'Citizenship and Democracy in Germany: Implications for Understanding Globalisation', in Vandenberg, A. (ed.), *Citizenship and Democracy in a Global Era* (London, Macmillan Press).

Kritsiotis, D. (2000), 'Imagining the International Community', *EJIL* 13, 961–92.

Kuratorium für einen demokratisch verfaßten Bund deutscher Länder (1991), *Vom Grundgesetz zur deutschen Verfassung. Denkschrift und Verfassungsentwurf* (Baden-Baden, Nomos).

Kuster, H. (1998), *Geschichte des Waldes: Von der Urzeit bis zur Gegenwart* (Munich, Beck).

Kymlicka, W. and Norman, W. (1994), 'Return of the Citizen: A Survey of Recent Work on Citizenship Theory', *Ethics* 104, 352–81.

Lang, W. (1999), 'UN Principles and International Environmental Law', *Max Planck Yearbook of United Nations Law* 3, 157–72.

Lazarus, R.J. (1986), 'Changing Conceptions of Property and Sovereignty on Natural Resources: Questioning the Public Trust Doctrine', *Iowa Law Review* 71, 631–716.

Lee, J. (2000), 'The Underlying Legal Theory to Support a Well-Defined Human Right to a Healthy Environment as a Principle of Customary International Law', *Columbia Journal of Environmental Law* 25, 283–346.

Leopold, A. (1949), *A Sand County Almanac* (Oxford, Oxford University Press).

Lodefalk, M. and Whalley, J. (2002), 'Reviewing Proposals for a World Environment Organisation', *The World Economy* 25:5, 601–17.

Low, N. and Gleeson, B. (1998), *Justice, Society and Nature* (London, Routledge).

Lowe, V. (1999), 'Sustainable Development and Unsustainable Arguments', in Boyle, A. and Freestone, D. (eds), *International Law and Sustainable Development: Past Achievements and Future Challenges* (Oxford, Oxford University Press).

Lowry, M.L. (1996), *Of Mice and Genes: Ethics and European Patent Law on Biotechnological Inventions* (Florence, European University Institute).

MacCormick, N. (1978), *Legal Reasoning and Legal Theory* (repr. 1994, Oxford, Clarendon Press).

MacCormick, N. (1993), 'Beyond the Sovereign State', *Modern Law Review* 56, 1–18.

MacIntyre, A. (1981), *After Virtue; A Study in Moral Theory* (London, Duckworth).

Mackey, B. (2004), 'The Earth Charter and Ecological Integrity', *Worldviews* 8, 76–92.

Macroy, R. (2004), 'Principles into Practice', in Macroy, R. (ed.), *Principles of European Environmental Law* (Groningen, Europa Law Publishing).

Marquardt, B. (2003), *Umwelt und Recht in Mitteleuropa: Von den großen Rodungen des Hochmittelalters bis ins 21.Jahrhundert* (Zurich, Schulthess).

Marquardt, B. (2005), 'Zeitenwende fur die Nachhaltigkeit: Zur umwelthistorischen Zasur um 1800', *GAIA* 14:3, 243–52(10).

May, P., Burby, R.J., Ericksen, N.J., Handmer, J.W., Dixon, J.E. and Michaels, S. (eds) (1996), *Environmental Management and Governance: Intergovernmental Approaches to Hazards and Sustainability* (London and New York, Routledge).

McIntosh, R. (1985), *The Background of Ecology: Concept and Theory* (Cambridge, Cambridge University Press).

Meadows, D.H., Meadows, D.L. and Randers, J. (1992), *Beyond the Limits: Confronting Global Collapse, Envisioning a Sustainable Future* (Vermont, Chelsea Green Publishing).

Meadows, D.H., Meadows, D.L., Randers, J. and Behrens, W. III (1972), *The Limits to Growth* (New York, Universe Books).

Meron, T. (1995), 'Extraterritoriality of Human Rights Treaties', *American Journal of International Law* 89, 78–82.

Mesarovic, M. and Pestel, E. (1972), *Menschheit am Wendepunkt* (Stuttgart, Deutsche Verlagsanstalt).

Mesarovic, M. and Pestel, E. (1983), 'Organisches und dauerhaftes Wachstum', in Peccei, A., Pestel, E. and Mesarovic, M. (eds), *Der Weg ins 21.Jahrhundert. Alternative Strategien fur die Industriegesellschaaft, Berichte an den Club of Rome* (Munich, Molden).

Mitcham, C. (1996), in Gottlieb, R. (ed.), *The Ecological Community* (New York and London, Routledge).

Mollo, M., Johl, A., Wagner, M., Popovic, N., Lador, Y. and Hoenninger, J. (2005), *Environmental Human Rights Report: Human Rights and the Environment – Materials for the 61st Session of the United Nations Commission on Human Rights, Geneva, March 14–April 22, 2005* (Oakland, CA, Earthjustice Legal Defense Fund).

Moltke, K. von (2005), 'Clustering International Environmental Agreements as an Alternative to a World Environment Organization', in Biermann, F. and Bauer, S. (eds) (2005), *A World Environment Organization* (Aldershot, Ashgate).

Morse, B. (2008), 'Indigenous Rights as a Mechanism to Promote Environmental Sustainability', in Westra, L., Bosselmann, K. and Westra, R. (eds), *Reconciling Human Existence and Ecological Integrity* (London, Earthscan).

Moser, W.G. (1757), *Grundsätze der Forst-Ökonomie* (Frankfurt and Leipzig).

Munansinghe, M. and Lutz, E. (1991), *Environmental-Economic Evaluation of Projects and Policies for Sustainable Development*, Environment Working Paper No. 42 (Washington DC, The World Bank).

Murcott, S. (1997), 'Appendix A: Definitions of Sustainable Development' (16 February), <www.aocweb.org/emr/Portals/2/MIT%20Definitions.pdf>, accessed 1 December 2007.

Murswiek, D. (2002), 'Nachhaltigkeit – Probleme der rechtlichen Umsetzung eines umweltpolitischen Leitbildes', *Natur und Recht*, 641–45.

Najam, A. (2005), 'Neither Necessary, Nor Sufficient: Why Organizational Tinkering Will Not Improve Environmental Governance', in Biermann, F. and Bauer, S. (eds) (2005), *A World Environment Organization* (Aldershot, Ashgate).

Nash, J.A. (1993), 'The Case for Biotic Rights', *Yale Journal of International Law* 18, 235–49.

New Zealand Ministry for the Environment (1989), *Resource Management Law Reform: Sustainability, Intrinsic Values and the Needs of Future Generations*, Working Paper No.24 (Wellington).

Newell, P. (2002), 'A World Environment Organisation: the Wrong Solution to the Wrong Problem', *The World Economy* 25, 659–71.

Nickel, J.W. (1993), 'The Human Right to a Safe Environment: Philosophical Perspectives on Its Scope and Justification', *Yale Journal of International Law* 18, 281–95.

Nolan, D. (2005), *Environmental and Resource Management Law*, 3rd edn (Wellington, LexisNexis).

Nordhaus, W. (1993), 'Economic Growth on a Planet under Siege', in Siebert, H. (ed.), *Economic Growth in the World Economy* (Tübingen, Mohr).

Odendahl, K. (1998), Die Umweltpflichtigkeit der Souveranitat: Reichweite und Schranken territorialer Souveranitatsrechte uber die Umwelt und die Notwendigkeit eines veranderten Verstandnisses staatlicher Souveranitat (Berlin, Duncker & Humblot).

Office of the Parliamentary Commissioner for the Environment (1998), *Towards Sustainable Development. The Role of the Resource Management Act 1991* (Wellington).

Office of the Parliamentary Commissioner for the Environment (2002), *Creating Our Future. Sustainable Development for New Zealand* (Wellington).

Office of the Parliamentary Commissioner for the Environment (2007), *Sustainability Review 2007: New Zealand's Progress Towards Sustainable Development* (Wellington).

Ostrom, E. (1990), *Governing the Global Commons: The Evolution of Institutions for Collective Action* (Cambridge, Cambridge University Press).

Ott, K. (2008), 'Solving the Demarcation Problem', in Westra, L., Bosselmann, K. and Westra, R. (eds), *Reconciling Human Existence and Ecological Integrity* (London, Earthscan).

Ottonelli, V. (2002), 'Immigration: What does Global Justice Require?', in Dower, N. and Williams, J. (eds), *Global Citizenship: A Critical Reader* (Edinburgh, Edinburgh University Press).

Pacific Institute of Resource Management (ed.) (1992), 'Commitment for the Future: The Earth Charter and Treaties Agreed to by the International NGOs and Social Movements', Paper presented to the International NGO and Social Movements Forum Conference, Wellington, New Zealand, 11 June.

Pallemaert, M. and Azmanova, A. (eds) (2006), *The European Union and Sustainable Development: Internal and External Dimensions* (Amsterdam, Kluwer).

Palmer, G. (1989), Verbatim Record of the 15th Meeting GA, 44th Session, 61, UN Doc. A/44/PV.15.

Palmer, G. (1992), 'New Ways To Make International Environmental Law', *American Journal of International Law* 86:2, 259–83.

Palmer, G. (1995), *Environment: The International Challenge* (Wellington, Victoria University Press).

Palmer, J. (1995), 'Just Ecological Principles?', in Cooper, D. and Palmer, J. (eds), *Just Environments – Intergenerational, International and Inter-Species Issues* (London, Routledge).

Pardy, B. (1997), 'Planning for Serfdom: Resource Management and the Rule of Law', *New Zealand Law Journal*, 69–72.

Pearce, D. (1993), *Blueprint 3: Measuring Sustainable Development* (London, Earthscan).

Pearce, D., Markandya, A. and Barbier, E.B. (1989), *Blueprint for a Green Economy* (London, Earthscan).

Pernthaler, P. (1992), 'Reform der Bundesverfassung', in Pernthaler, P., Weber, K. and Wimmer, N. (eds), *Umweltpolitik durch Recht-Moglichkeiten und Grenzen* (Wien, Manz).

Pezzey, J. (1992), 'Sustainability: An Interdisciplinary Guide', *Environmental Values* 1, 321–62.

Picht, G. (1971), 'Umweltschutz und Politik', *Zeitschrift für Rechtspolitik* 4, 152–8.

Politis, N. (1925/I), 'Le Probleme des limitations de la souverainete et la theorie de l'abus des droits dans les rapports internationaux', *Recueil des Cours à l'Académie de Droit International* 6, 1–121.

Porter, G. and Brown, J.W. (1996), *Global Environmental Politics*, 2nd edn (Boulder CO, Westview Press).

Postiglione, A. (2001), 'An International Court for the Environment', in Low, N. and Gleeson, B. (eds), *Governance for the Environment: Global Problems, Ethics and Democracy* (London, Palgrave).

Praetorius, I. and Saladin, P. (1994), *Die Würde der Kreatur*, Schriftenreihe Umwelt Nr. 260, BUWAL (ed.), 121.

Pye-Smith, C. and Feyerabend, G.B. (1994), *The Wealth of Communities: Stories of Success in Environment Management* (London, Earthscan).

Radkau, J. (2000), *Natur und Macht. Eine Weltgeschichte der Umwelt* (Munich, Beck).

Raftopoulos, E. (1992), 'The Barcelona Convention System for the Protection of the Mediterranean Sea against Pollution: An International Trust at Work', *International Journal of Estuarine and Coastal Law* 7, 27–41.

Rawls, J. (1993), *A Theory of Justice*, rev. edn (New York, Oxford University Press).

Rawls, J. (1993), *Political Liberalism* (New York, Oxford University Press).

Raz, J. (1975), *Practical Reason and Norms*, repr. 1990 (London, Hutchinson).

Raz, J. (1986), *The Morality of Freedom* (Oxford, Oxford University Press).

Redgwell, C. (1996), 'Life, The Universe and Everything: A Critique of Anthropocentric Rights', in Boyle, A. and Anderson, M.R. (eds), *Human Rights Approaches to Environmental Protection* (New York, Clarendon Press).

Regan, T. (1992), 'Does Environmental Ethics Rest on a Mistake?', *Monist* 75, 161–82.

Repetto, R. (1986), *World Enough and Time* (New Haven, Yale University Press).

Rieser, A. (1991), 'Ecological Preservation as a Public Property Right: An Emerging Doctrine in Search of a Theory', *Harvard Environmental Law Review* 15, 393–433.

Riphagen, W. (1980), 'The International Concern for the Environment as Expressed in the Concepts of "Common Heritage" and of "Shared Natural Resources"', in Bothe, M. (ed.), *Trends in Environmental Law and Policy* (Berlin, Beitrage zur Umweltgestaltung).

Robb, C. (ed.) (2001), *International Environmental Law Reports: Trade and Environment* (Cambridge, Cambridge University Press), vol. 2.

Robb, C. (ed.) (2001), *International Environmental Law Reports: Trade and Environment* (Cambridge, Cambridge University Press), vol. 3.

Robinson, N. (1997), *IUCN (The World Conservation Union) Newsletter*, July–September.

Rockefeller, S. (2005), in Blaze Corcoran, P. (ed.), *The Earth Charter in Action. Toward a Sustainable World* (Amsterdam, KIT).

Rolston, H. III (1993), 'Rights and Responsibilities on the Home Planet', *Yale Journal of International Law* 18. 251–79.

Ruttan, V. (1988), 'Sustainability is Not Enough', *AJAA* 2, 607.

Sadeleer, N. de (2002), *Environmental Principles* (Oxford, Oxford University Press).

Sadeleer, N. de (2004), 'Environmental Principles, Modern and Post-Modern Law', in Macroy, R. (ed.), *Principles of European Environmental Law* (Groningen, Europa Law Publishing).

Saladin, P. and Schweizer, R.J. (1995), *Kommentar zur Bundesverfassung der Schweizerischen Eidgenossenschaft*, Art. 24 novies Abs. 3 (Basle, Verlag Helbing & Lichtenhahn).

Salmon, P. (2002), 'Sustainable Development in New Zealand', Paper presented to the Auckland Branch Resource Management Law Association on 30 October.

Sand, P. (2002), 'Trusteeship for Common Pool Resources: Zur Renaissance des Treuhandbegriffs im Umweltvölkerrecht', in Schorlemer, S. (ed.), *Praxishandbuch UNO* (Berlin, Springer).

Sand, P. (2004), 'Sovereignty Bounded: Public Trusteeship for Common Pool Resources?', *Global Environmental Politics* 4, 47–71.

Sand, P. (2006), 'Global Environmental Change and the Nation State', in Winter, G. (ed.), *Multilevel Governance of Global Environmental Change* (Cambridge, Cambridge University Press).

Sand, P. (2007), 'Public Trusteeship for the Oceans', in Wolfrum, R., Ndiaye, T.M. and Kojima, C. (eds), *Law of the Sea, Environmental Law and Settlement of Disputes: Liber Amicorum Judge Thomas A. Mensah* (Lieden, Martinus Nijhoff).

Sands, P. (2003), *Principles of International Environmental Law*, 2nd edn (Cambridge, Cambridge University Press).

Sax, J.L. (1970), 'The Public Trust Doctrine in Natural Resources Law: Effective Judicial Intervention', *Michigan Law Review* 68, 471–556.

Sax, J.L. (1980), 'Liberating the Public Trust Doctrine from its Historic Shackles', *UC Davis Law Review* 14, 185–94.

Schanz, H. (1996),'Forstliche Nachhaltigkeit. Sozialwissenschaftliche Analyse der Begriffsinhalte und – funktionen', Ph.D. diss.(University of Freiburg).

Schlosberg, D. (2001), 'Three Dimensions of Ecological Justice', Paper prepared for Political Research Annual Joint Sessions, Grenoble, France, 6–11 April (updated 10 October 2004), <http://www.essex.ac.uk/ecpr/events/jointsessions/paperarchive/grenoble/ws6/schlosberg.pdf> (draft), accessed 1 December 2007.

Schlosberg, D. (2003), 'The Justice of Environmental Justice: Reconciling Equity, Recognition, and Participation in a Political Movement', in Light, A. and De-Shalit, A. (eds), *Moral and Political Reasoning in Environmental Practice* (Cambridge, MA, MIT Press).

Schlosberg, D. (2005), 'Environmental and Ecological Justice: Theory and Practice in the US', in Eckersley, R. and Barry, J. (eds), *The State and the Global Ecological Crisis* (Cambridge, MA, MIT).

Schlosberg, D. (2007), *Defining Environmental Justice* (Oxford, Oxford University Press).

Schmidt, R. and Muller, H. (2001), *Einfuhrung in das Umweltrecht*, 6th edn (Munich, Beck).

Schroter, M. (1999), *Mensch, Erde, Recht. Grundfragen okologischer Rechtstheorie* (Baden-Baden, Nomos).

Schwarz, E., '*Oberlandjagermeister v. Gochhausen*', *Archiv fur Forstwesen* 9:7.

Shelton, D. (1991), 'Human Rights, Environmental Rights, and the Right to Environment', *Stanford Journal of International Law* 28, 103–38.

Shrijver, N. (1989), 'International Organization for Environmental Security', *Bulletin of Peace Proposals* 20:2, 115–22.

Shrijver, N. (1997), *Sovereignty over Natural Resources: Balancing Rights and Duties* (Cambridge, Cambridge University Press).

Sieferle, R.P. (1998), 'Wie tragisch war die Allmende?', *GAIA* 7:4, 304–7(4).

Simma B. (1993), 'Does the UN-Charter Provide an Adequate Legal Basis for Individual or Collective Responses to Violations of Obligations erga omnes?', in Delbrück, J. (ed.), *The Future of International Law Enforcement* (Berlin, Duncker & Humblot).

Simms, A. (2005), 'Economy: The Economic Problem of Sustainable Governance', in Ayre, G. and Callway, R. (eds), *Governance for Sustainable Development: A Foundation for the Future* (London, Earthscan).

Simonis, U. (2002), 'Advancing the Debate on a World Environment Organisation', *The Environmentalist* 22:1, 29–42.

Singer, P. (1975), *Animal Liberation: A New Ethics for our Treatment of Animals* (New York, Random House).

Singer, P. (2003), 'Animal Liberation at 30', *New York Times Review of Books*, Vol. 50, No. 8.

Sixth Environment Action Programme. Environment 2010: Our Future, Our Choice, <http://europa.eu.int/comm/environment/newprg>, accessed 1 December 2007.

Skelton, P. and Memon, A. (2002), 'Adopting Sustainability as an Overarching Environmental Policy: A Review of Section 5 of the RMA', *Resource Management Journal* 10:1, March, 8–9.

Sloane, R.D. (2001), 'Outrelativizing Relativism: A Liberal Defense of the Universality of International Human Rights', *Vanderbilt Journal of Transnational Law* 34, 527–95.

Smith, J. and Shearman, D. (2006), *Climate Change Litigation* (Adelaide, Presidian Legal Publications).

Steinberg, R. (1998), *Der okologische Verfassungsstaat* (Frankfurt, Suhrkamp).

Steenbergen, B. van (1994), 'The Condition of Citizenship: An Introduction', in Steenbergen, B. van (ed.), *The Condition of Citizenship* (London, Sage Publications).

Steenbergen, B. van (1994), 'Towards a Global Ecological Citizen', in Steenbergen, B. van (ed.), *The Condition of Citizenship* (London, Sage Publications).

Steiner, H.J., Alston, P. and Goodman, R. (eds) (2007), *International Human Rights in Context* (Oxford, Clarendon Press).

Steurer, R. (2001), 'Paradigmen der Nachhaltigkeit', *Zeitschrift für Umweltrecht und Umweltpolitik* 4, 537–39.

Stocker, W. (1993), *Das Prinzip des Common Heritage als Ausdruck des Staatengemeinschaftsinteresses im Volkerrecht* (Zurich, Schulthess Juristische Medien).

Stone, C.D. (1972), 'Should Trees Have Standing?', *South California Law Review* 45, 450–501.

Stone, C.D. (1974), *Should Trees Have Standing?: Toward Legal Rights for Natural Objects* (Los Altos, Kaufmann).

Stone, C.D. (1987), *Earth and Other Ethics: The Case for Moral Pluralism* (New York, Harper and Row).

Stone, C.D. (1993), *The Gnat is Older than Man: Global Environment and Human Agenda* (Princeton, Princeton University Press).

Stone, C.D. (1996), *Should Trees Have Standing? And other Essays on Law, Morals and the Environment: 25th Anniversary Edition* (New York, Oxford University Press).

Stone, C.D. (2001), 'Is there a Precautionary Principle', *Environmental Law Reporter* 31, 10790.

Strong, M. (1989), 'The United Nations in an Interdependent World', *International Affairs*, January, 11–21.

Strong, M. (2005), 'A People's Earth Charter', in Blaze Corcoran, P. et al. (eds), *Toward a Sustainable World: The Earth Charter in Action* (The Hague, Kluwer International).

Suter, K. (1991), *Antarctica: Private Property or Public Heritage?* (London, Pluto Press).

Tainter, J.A. (1988), *The Collapse of Complex Societies* (Cambridge, Cambridge University Press).

Tainter, J.A. (2000), 'Problem Solving – Complexity, History, Sustainability', *Population and Environment* 22:3, 3–41.

Tarlock, A.D. (1997), 'Exclusive Sovereignty versus Sustainable Development of a Shared Resource: The Dilemma of Latin American Rainforest Management', *Texas International Law Journal* 32, 37–66.

Taylor, P. (1998), *An Ecological Approach to International Law* (London and New York, Routledge).

Taylor, P. (1998), 'From Environmental to Ecological Human Rights: A New Dynamic in International Law?', *Georgetown International Environmental Law Review* 10, 309–97, 384–92.

Taylor, P. (2002), 'The Global Perspective: Convergence of International and Municipal Law', in Bosselmann, K. and Grinlinton, D. (eds), *Environmental Law for a Sustainable Society* (Auckland, New Zealand Centre for Environmental law), vol.1.

Taylor, P. (2003), 'Reforming the Governance of the European Union: A Greater Voice and Expanded Planning Role for Local Government in EU Affairs?', in Lilly, I. and Bosselmann, K. (eds), *Repositioning Europe: Perspectives from New Zealand*, NCRE Research Series No.2 (Christchurch, University of Canterbury).

Taylor, P. (2007), 'Climate Change Litigation: A Catalyst for Corporate Responses' in *Climate Action*, published by Sustainable Development International in Partnership with United Nations Environment Program (London, Henley Media Group), 104–7.

Taylor, P. (2008), 'Ecological Integrity and Human Rights', in Westra, L., Bosselmann, K. and Westra, R. (eds), *Reconciling Human Existence and Ecological Integrity* (London, Earthscan).

Taylor, P. and Bosselmann, K. (2007), 'The Earth Charter in the Classroom: Transforming the Role of Law', in *Education for Sustainable Development in Action – Good Practices Using the Earth Charter* (San José, Costa Rica, UNESCO and Earth Charter International).

Theutenberg, B. (1984), 'The International Environmental Law – Some Basic Viewpoints', *Acad Droit Colloq*, 233.

Tladi, D. (2002), 'Of Course for Humans: A Contextual Defence of Intergenerational Equity', *South African Journal of Environmental Law and Policy* 9, 177–92.

Tladi, D. (2007), *Sustainable Development in International Law* (Pretoria, Pretoria University Press).

Tomashow, M. (1995), *Ecological Identity: Becoming a Reflective Environmentalist* (Cambridge MA, MIT).

Tronto, J. (1993), *Moral Boundaries: A Political Argument for the Ethic of Care* (London and New York, Routledge).

Tully, J. (1995), *Strange Multiplicity: Constitutionalism in an Age of Diversity* (Cambridge, Cambridge University Press).

UNESCO (2003), 'UNESCO's Support for the Earth Charter' in UNESCO, *Records of the General Conference 32nd session Paris, 29 September to 17 October 2003* (Paris, UNESCO), vol.1, 35.

UNESCO, 'Education for Sustainable Development: Highlights on DESD Progress to Date' (April 2007), <www.portal.unesco.org/education/en/files/51172/11779357975Progress_to_Date_APRIL07.pdf/Progree+to+Date+APRIL07.pdf>, accessed 1 December 2007.

UNESCOR Commission on Human Rights, Sub-Commission on Prevention of Discrimination and Protection of Minorities (1994), *Review of Further Developments in Fields with Which the Sub-Commission Has Been Concerned on Human Rights and the Environment: Final Report Prepared by Mrs. Fatma Zohra Ksentini, Special Rapporteur*, UN Doc. E/CN.4/Sub.2/1994/9.

Upton, S. (1994), 'The Resource Management Act, Section 5: Sustainable Management of Natural and Physical Resources', Keynote Speech to the Second Annual Conference of the Resource Management Law Association, Wellington, New Zealand, October.

Upton, S., Atkins, H. and Willis, G. (2002), 'Section 5 Re-visited; A Critique of Skelton and Memon's Analysis', *Resource Management Journal* 10:3, November, 10–22.

Urmson, J. (1988), *Aristotle's Ethics* (Oxford, Oxford University Press).

Van Dyke, J.M. (1993), 'International Governance and Stewardship of the High Seas and Its Resources', in Van Dyke, J.M., Zaelke, D. and Hewison, G. (eds), *Freedom for the Seas in the 21st Century: Ocean Governance and Environmental Harmony* (Washington, DC, Island Press).

Verschuuren, J. (2003), *Principles of Environmental Law* (Baden-Baden, Nomos).

Vilela, M. (2005), 'Building Consensus on Shared Values. History and Provenance of the Earth Charter', in Blaze Corcoran, P. (ed.), *The Earth Charter in Action. Toward a Sustainable World* (Amsterdam, KIT).

Waller, M. and Linklater, A. (2003), *Political Loyalty and the Nation-State* (London and New York, Routledge).

Weidner, H., Jänicke, M. and Jorgens H. (eds) (2002), *Capacity Building in National Environmental Policy: A Comparative Study of 17 Countries* (Heidelberg, Springer Verlag), <http://www.pce.govt.nz/projects/susstrategy.pdf >, accessed 1 December 2007.

Weil, P. (1983), 'Towards Relative Normativity in International Law?', *American Journal of International Law* 77, 413–42.

Westra, L. (2007), *Environmental Justice and the Rights of Indigenous Peoples* (London, Earthscan).

Westra, L. Bosselmann, K. and Westra, R. (eds) (2008), *Reconciling Human Existence and Ecological Integrity: Science, Ethics, Economics and Law* (London, Earthscan).

Wilkinson, C.F. (1989), 'The Headwaters of the Public Trust: Some Thoughts on the Source and Scope of the Traditional Doctrine', *Environmental Law* 19, 425–72.

Williams, D.A.R. (1997), *Environmental and Resource Management Law* (Wellington, Butterworths).

Winter, G. (2004), 'The Legal Nature of Environmental Principles in International, EC and German Law', in Macroy, R. (ed.), *Principles of European Environmental Law* (Groningen, Europa Law Publishing).

Winter, G. (2006), 'Introduction', in Winter, G. (ed.), *Multilevel Governance and Global Environmental Change* (Cambridge, Cambridge University Press).

Wissenburg, M. (1999), 'An Extension of the Rawlsian Savings Principle to Liberal Theories of Justice in General', in Dobson, A. (ed.), *Fairness and Futurity: Essays on Environmental Sustainability and Environmental Justice* (Oxford, Oxford University Press).

Wolfrum, R. (1990), 'Purposes and Principles of International Environmental Law', *German Yearbook of International Law*, 308–30.

World Bank (1992), *World Development Report 1992: Development and the Environment* (New York, Oxford University Press).

Wynter, M. (1998), 'The Compatibility of Trade and Environmental Concerns: Lessons From the Shrimp-Turtle Dispute', in Anghie, A. and Sturgess, G. (eds), *Legal Visions of the 21st Century: Essays in Honour of Judge Christopher Weeramantry* (The Hague and Boston, Kluwer Law International).

Young, I. (1990), *Justice and the Politics of Difference* (Princeton, Princeton University Press).

Young, O. (1982), *Resource Regimes: Natural Resources and Social Institutions* (Berkeley, University of California Press).

Young, O. (ed.) (1997), *Global Governance: Drawing Insights from the Environmental Experience* (Cambridge, MA, MIT).

Yves Lador 'Time for a Universal Declaration on Environmental Rights', <http://partnerships4planet.ch/en/environmental-rights.php>.

Zahren, W.M. von (1998), 'Ocean Ecosystem Stewardship', *William and Mary Environmental Law and Policy Review* 23, 108–20.

International Treaties

Additional Protocol to the American Convention on Human Rights in the Area of Economic, Social and Cultural Rights (Protocol of San Salvador) (San Salvador, El Salvador, 17 November 1988).

Amazonian Treaty 1978.

Amsterdam Treaty 1997.

Antarctic Treaty 1959 (Washington, 1 December 1959; 402 United Nations Treaty Series, 71).

Antigua Convention 1992.

ASEAN Convention 1985.

Biodiversity Convention 1992.

Convention for the Protection of the World Cultural and Natural Heritage (16 November 1972; 11 ILM 1358).

Convention for the Regulation of Whaling 1946.

Convention on Biological Diversity 1992 (Rio de Janiero, 5 June 1992; 31 ILM 818 (1992)).
Convention on International Trade in Endangered Species of Wild Fauna and Flora 1973.
Convention on Long-Range Transboundary Air Pollution (Geneva, 13 November 1979; 18 ILM 1449).
Convention on the Conservation of Antarctic Marine Living Resources (Canberra Convention) (Canberra, 20 May 1980; 19 ILM 841).
Convention on the Conservation of European Wildlife and Natural Habitats (Bern Convention) (Bern, 19 September 1979; European Treaty Series 104).
Convention on the Elimination of All Forms of Discrimination Against Women 1979.
Convention on the Rights of the Child 1989.
Desertification Convention 1997.
General Agreement on Trade in Services 1994.
International Covenant on Civil and Political Rights 1966.
International Covenant on Economic, Social and Cultural Rights 1966.
International Labor Convention 1949.
Kuwait Convention 1978.
Kyoto Protocol 1997.
Maastricht Treaty 1992.
Marrakesh Agreement Establishing the World Trade Organization 1994.
Moon Treaty 1957.
Paris Convention 1974.
Protocol on Environmental Protection to the Antarctic Treaty 1991 (Madrid Protocol) (Madrid, 4 October1991; 30 ILM 1461).
Ramsar Convention on Wetlands 1971.
Statute of the International Court of Justice 1946.
Tokyo International Convention for the High Seas Fisheries of the Northern Pacific Ocean 1952.
Treaty Establishing a Constitution for Europe 2004 (Doc. 2004/C310/01).
Treaty of Westphalia 1648.
UNESCO Convention for the Protection of the World Cultural and Natural Heritage (Paris, 23 November 1972; 11 ILM. 251).
United Nations Convention on the Law of the Sea 1982.
United Nations Convention to Combat Desertification 1994.
United Nations Framework Climate Change Convention 1992 (New York, 9 May 1992; UN Doc. A/AC.237/18, 31 ILM 848).
Universal Declaration of Human Rights 1949.

Charters, Resolutions, Reports

African Charter on Human and Peoples' Rights 1981 (Banjul Charter).
Agenda 21: Programme of Action for Sustainable Development (Agenda 21) (Rio de Janeiro, 14 June 1992; UN Doc. A/Conf.151/26 (1992), 31 ILM 874 (1992)).

Agenda item 148 for the United Nations General Assembly, 'Conservation of the climate as part of the common heritage of mankind' (30 November 1988; UN Doc. A/43/905).

Charter of Economic Rights and Duties of States, UN General Assembly Resolution 3281 (XXIX) of 12 December 1974 (ILM 14 (1975)).

Charter of Fundamental Rights of the European Union, adopted 7 December 2000, 2000/C 364/01, <http://www.europarl.europa.eu/charter/pdf/text_en.pdf>.

Charter of the United Nations (San Francisco, 26 June 1945).

Commission Proposal for a Directive on Access to Justice (03/0624 final COD 2003/0246).

Declaration of the United Nations Conference on the Human Environment (Stockholm Declaration) (Stockholm, 16 June 1972; UN Doc. A/Conf.48/14, 11 ILM 1461 (1972)).

Declaration on the Right to Development 1986 (4 December 1986; UN Doc. A/41/53).

Directive 2003/4/EEC (28 January 2003), Public Access to Environmental Information.

Directive 2003/35/EEC (26 May 2003), Providing Public Participation in Respect of the Drawing Up of Certain Plans and Programmes relating to the Environment.

Draft Article on State Responsibility 1996 (UN Doc. A/CN.4/L.528/Add.2).

Draft International Covenant on Environment and Development (March 1995; World Conservation Union (IUCN), Environmental Policy and Law Paper No. 31, Rev.2).

Earth Charter, <www.earthcharter.org>.

Environment Charter ('Charte de l'environnement'), <http://www.ecologie.gouv. fr/IMG/pdf/affiche_charte_environnement.pdf>.

ILC Report 1989 (UN Doc. A/CN 4/423), 22 (54).

ILC Report 1993 (UN Doc. A/CN.4/450).

Implementation of a Regime of Equal Right of Access and Non-Discrimination in Relation to Transfrontier Pollution, OECD (Recommendation C (77) 28 (1977)).

International Union for the Conservation of Nature and Natural Resources *Endorsement of the Earth Charter* (Res. WCC 3.022 (2004)), <www.iucn.org/congress/resolutions>.

International Union for the Conservation of Nature Draft International Covenant on Environment and Development 1995.

Johannesburg Declaration on Sustainable Development (Johannesburg Declaration) (Johannesburg, 4 September 2002, UN Doc. A/Conf.199/20 (2002)).

Johannesburg Plan of Implementation (Johannesburg, 4 September 2002, UN Doc. A/Conf.199/20).

New Delhi Declaration on the Principles of International Law Related to Sustainable Development (London, 2002; ILA resolution 3/2002).

OAS Hemispheric Summit on Sustainable Development 1997.

OECD Council Recommendation C (974) 224 of 14 November 1974.

Principles Concerning Transfrontier Pollution, OECD (Recommendation C (74) 224, 14 November 1974).

Report of the Secretary-General to the General Assembly, 'Environment and Human Settlements' (6 October 1998; UN Doc. A/53/463).

Report of the Secretary-General to the General Assembly, 'Renewing the United Nations: a programme for reform' (14 July 1997; UN Doc. A/51/950).

Resolution on a Draft Constitution of the European Union 1994, OJC 61/155.

Rio Declaration on Environment and Development (the Rio Declaration) (Rio de Janeiro, 13 June 1992; UN Doc. A/Conf.151/26 (vol. I), 31 ILM 874 (1992)).

United Nations Conference on Environment and Development (Rio de Janiero, 3–14 June 1992; 31 ILM 814).

United Nations Conference on Population and Development 1994.

United Nations Declaration on the Rights of Indigenous Peoples, A/RES/61/295, 13 September 2007.

United Nations Environmental Programme Governing Council, seventh special session (Cartagena Declaration) (Cartagena, Colombia, 15 February 2002; Decision SS.VII/1).

United Nations General Assembly Resolution 217A (III) of 10 December 1948.

United Nations General Assembly Resolution 523 (VI) of 12 January 1952, United Nations Yearbook 5 (1952).

United Nations General Assembly Resolution 626 (VII) of 21 December 1952, United Nations Yearbook 6 (1952).

United Nations General Assembly Resolution 837 (IX) of 14 December 1954, United Nations Yearbook 12 (1958).

United Nations General Assembly Resolution 3129 (XXVIII) of 13 December 1973.

United Nations General Assembly Resolution 37/7 of 28 October 1982.

United Nations General Assembly Resolution 45/94, UN Doc. A/RES/45/94 (1990), 'Need to Ensure a Healthy Environment for the Well-Being of Individuals'.

United Nations General Assembly Resolution 60/1 of 16 September 2005.

United Nations Millennium Declaration (GA Res. 55/2, UN GAOR, 55th session, 8th plenary meeting, UN Doc. A/Res/55/2 (2000)).

Universal Declaration of Human Rights (UN Doc. A/810 (1948)).

World Charter for Nature 1982 (GA Res. 37/7, UN GAOR, 37th session, 48th plenary meeting, UN Doc. A/37/7 (1983)).

World Commission on Environment and Development (1987), *Our Common Future*, 'Brundtland Report' (Oxford and New York, Oxford University Press).

World Summit for Social Development 1995.

Case Law

Arrondelle *v.* United Kingdom, App. No. 7889/77, 5 Eur. HR Rep. 118, 119 (1982) (European Commission on Human Rights) (friendly settlement).

Case Concerning Maritime Delimitation in the Area between Greenland and Jan Mayen (Denmark/Norway), ICJ 38; 99 ILR 395, (1993).

Case Concerning the Gabçikovo-Nagmaros Project (Hungary/Slovakia), 1997 ICJ; 37 ILM 162 (1998), Separate Opinion of Vice-President Weeramantry.

Corfu Channel Case (United Kingdom of Great Britain and Northern Ireland/ Albania), ICJ Reports 1949.

EHP *v.* Canada, Communication No. 67/1980, in United Nations Human Rights Committee (1990), *Selected Decisions of the Human Rights Committee under the Optional Protocol* (United Nations Publications), vol. 2, 20.

Gentini Case (Italy/Venezuela) MCC (1903).

Lopez Ostra *v.* Spain, judgment of 9 December 1994, Series A no. 303-C, § 51.

New Zealand Rail Ltd *v.* Marlborough District Council [1994] NZRMA 70.

North Shore City *v.* Auckland Regional Council [1997] NZRMA 59.

Ominayak and the Lubicon Lake Band *v.* Canada, UN Human Rights Committee, (communication no. 167 (1984); UN Doc. CCPR/C/38/D/167/1984).

Oneryildiz *v.* Turkey [2004] ECHR 657 (30 November 2004).

Powell and Rayner *v.* United Kingdom, 172 Eur. Ct. HR (Series A), (1990).

Southern Bluefin Tuna Case, (Australia and New Zealand/ Japan), 38 ILM 1624 (1999).

Taskin and Others *v.* Turkey, 46117/99 [2004] ECHR 621 (10 November 2004).

Trail Smelter Arbitration United States *v.* Canada (1931–41; 3 Reports of International Arbitral Awards, 1938.

Trio Holdings *v.* Marlborough District Council [1997] NZRMA 97.

United States Restrictions on Imports of Tuna (European Comm. and Netherlands *v.* US) GATT Panel Report 16 June 1994, both reported in 2 IELR 48–114.

United States Restrictions on Imports of Tuna (Mexico *v.* US) GATT Panel Report 3 September 1991.

Yanomami Indians *v.* Brazil Inter-Am. (CHR 7615, OEA/Ser.L.V/II/66 doc. 10 rev. 1 (1985)), <http://www.cidh.org/annualrep/84.85eng/Brazil7615.htm>.

Statutes

Local Government Act 2002 (New Zealand).

National Environmental Protection Act 1969 (United States of America).

Resource Management Act 1991 (New Zealand).

Index

Community of life 38